Renate Barber

UNWIN UNIVERSITY BOOKS

UNWIN UNIVERSITY BOOKS

10

Income Distribution
and Social Change

A STUDY IN CRITICISM

RICHARD M. TITMUSS

Income
Distribution
and
Social Change

A STUDY IN CRITICISM

London
GEORGE ALLEN & UNWIN LTD
RUSKIN HOUSE   MUSEUM STREET

UNWIN UNIVERSITY BOOKS
George Allen & Unwin Ltd
40 Museum Street, London, WC1

PRINTED IN GREAT BRITAIN
in 10 point Times Roman type
BY SIMSON SHAND LTD
LONDON, HERTFORD AND HARLOW

To Ann

'We, Your Majesty's most dutiful and loyal subjects, the Commons of the United Kingdom in Parliament assembled, towards raising the necessary supplies to defray Your Majesty's public expenses and making an addition to the public revenue, have freely and voluntarily resolved to give and grant unto Your Majesty the several duties hereinafter mentioned.'

(preamble to the Annual Finance Act)

ACKNOWLEDGEMENTS

This book has been a long time in the making. Some of the difficulties are explained in the last chapter. It was both helped and hindered on its journey by advice and comments from friendly critics and critical friends. If I have not always heeded them it is not because I was brought up badly. Time was the enemy.

To Mr Tony Lynes, who acted for long spells as my research assistant and who has contributed two of the appendices, I owe a special word of thanks for his help in innumerable ways, not least for his shrewd and fearless comments on much that I wrote.

I wish also to record my gratitude to the following who read the manuscript in various stages and did so with patience and much constructive criticism: Dr Brian Abel-Smith, Mrs Margaret Gowing, Mr I. M. D. Little, Mr J. R. S. Revell, Mr Peter Townsend, Mr John Vaizey and Professor G. S. A. Wheatcroft. The conversations I had with them both before and after this process contributed in no small way to the development of this study. The opinions expressed and errors committed in the final outcome are not their responsibility. For technical assistance on a number of points I am grateful to Mr Geoffrey Smith and Mr Gordon Kirkwood. And for help in various ways and in mastering so much untidiness I would like to thank my secretary, Mrs Angela Vivian.

Elsewhere in the text I have recorded my appreciation of assistance received from officials in the Board of Inland Revenue in regard to a questionnaire I submitted. Acknowledgements are also due to the officials of the following bodies for permission to quote from evidence in their name given to the Royal Commission on the Taxation of Profits and Income: The Association of HM Inspectors of Taxes, The Inland Revenue Staff Federation, and The Association of Certified and Corporate Accountants.

The external costs of this book in family disservices are difficult to compute. Only Kay and Ann know the answer.

R.M.T.
Acton, December 1961

CONTENTS

CONTENTS

TABLES

TABLES

CHAPTER 1

Introduction

MUCH has been written in recent years about the strongly egalitarian effects of the social and economic policies pursued by British Governments since the end of the Second World War. The introduction of new social services and the expansion of existing ones contributed to this trend, it was argued, by redistributing resources in favour of wage-earners and the poorer income groups. Full employment, exploited by the trade unions for their own ends regardless of the national interests, was leading to a steady reduction in wage and salary differentials. The position of the rentier, adversely affected by rent control, inflation and other factors had, it was also believed, greatly deteriorated compared with pre-war days. The incomes of professional people and other self-employed middle-class groups, both before and after tax, were steadily losing ground in the socialist 1940's and the inflationary 1950's, relative to the gains made by wage-earners and other sections of the population who, before 1939, were at the bottom of the income scale. Above all, it was considered that the wealthy were a disappearing class in Britain. Not only had their incomes been sharply diminished by a variety of factors, one being the long-term and continuing effects of high estate duties, but, equally, high taxation was forcing them to spend their savings and reduce their standards of living.

Summed up in one general paragraph, this is what a great many people believed was happening to British society in the 1950's and early 1960's. To advocate seriously, as some responsible writers were doing in 1961, the complete abolition of all estate and inheritance duties was explicable on the premise that equality of incomes, before and after tax, had been pushed as far as was desirable in a predominantly private enterprise economy.[1] Nor, accepting this interpretation of contemporary trends in the distribution of incomes, was there anything inconsistent in the Government's decision, embodied in the National Insurance Acts of 1959–61 and the Finance Act of

---

[1] See, for example, *Crossbow*, April 1961, and Sewill, Brendon, *Treasure on Earth; A Study of Death Duties*, Conservative Political Centre, 1960. In *Principles and Practice*, a series of Bow Group essays for the 1960's, it was argued that 'our tax structure should be as non-progressive as possible' (p. 59).

1961, to shift the incidence of taxation from progressive to regressive and indirect systems of raising revenue, both central and local.

With qualifications and different shades of emphasis, these views were sincerely held and widely cultivated by many writers on social and economic topics during the late 1940's and throughout the 1950's. They found expression in a steady stream—at times, almost a torrent —of books, pamphlets, articles and newspaper editorials written for a wide audience ranging from the man-in-the-street to the economic specialist.[1] Most of them united in condemning the egalitarian effects of full employment, the reduction of differences in earnings and rewards for work, high progressive taxation, and the 'Welfare State'. For much of their evidence they drew liberally on the Annual Reports of the Board of Inland Revenue. The Board itself, in one of its post-war reports published in October 1950, made its one and only attempt to survey the changes in the distribution of incomes before and after tax between 1938–9 and 1948–9.[2] It concluded that there had been 'a very considerable redistribution in incomes'. This report, in providing both material and stimulus, was of profound importance in influencing much that was written on the subject of equality and incentives in subsequent years.

The official statistics of income also played a powerful role in shaping the recommendations of three critically important commissions and committees of the 1950's—the Royal Commissions on Doctors' and Dentists' Remuneration; on the Civil Service; and the Committee on the Taxation Treatment of Provisions for Retirement. The acceptance by the Government of the main recommendations of these bodies had a far-reaching effect on the whole structure of rewards for higher income groups, in the private sector as well as the public services. These reports, perhaps more than any others, are the keys to understanding some of the major characteristics of British society in the 1960's. In their formulation, the members, their secretariats, and many of the witnesses who gave evidence before them, accepted or drew heavily upon the official statistics on the size distribution of incomes. For most of their needs, there were no alternative and equally comprehensive data available.

We thus see that many of the powerful forces of change which go to the shaping of the economic and social structure depend, for their

---

[1] Some of the more influential and widely read books included: Lewis R. and Maude A., *The English Middle Classes*, 1949, and *Professional People*, 1952; De Jouvenel B., *The Ethics of Redistribution*, 1951; Rowntree S. and Lavers G. R., *Poverty and the Welfare State*, 1951; Dudley Seers, *The Levelling of Incomes since 1938*, 1951; Powell E., *Saving in a Free Society*, 1960.

[2] *BIR 92*, pp. 82–6. Further references to the annual reports of the Board will, for convenience, be similarly shortened. The Command Paper numbers are given in the index.

collective intelligence, on this particular source of information. It is rarely the function of these bodies to question in any fundamental way the validity of the statistics they use; moreover, these and other consumers normally do not have the time or the expert competence to do so. The role that the Webbs once played, as private critics of public statistics, is today less possible. Yet, at the same time, because of the increasing specialization and division of labour in society, the Blue Books on income and wealth are coming to play a more important part in the multitude of decisions which have to be made about rewards by committees rather than by individual employers. Committees demand and need facts, and they need them especially when more and more account has to be taken in the debate about equity of differences in rewards and benefits both inside and outside a particular occupation, profession or class.

Chancellors of the Exchequer also need more facts in the framing of social and fiscal policies as well as for the general oversight of the economy. What this means to the staff of the Board of Inland Revenue in the three months before the budget is vastly different from what it meant only a few decades ago. The days are long since past when policies for health, education and social security could be shaped without reference to trends in the distribution of income and wealth. Whatever opinions may be held as to the desirable role of these instruments of welfare, it is indisputable that in terms of finance they are inextricably mixed up with the state of the economy as a whole. On the interpretation then of the Board's statistics rests the use of taxation in its many forms as a weapon of social as well as economic policy.

The value of these statistics has increased immeasurably since the responsibilities of the Chancellor have no longer been confined to (what the 1944 *White Paper on Employment Policy* described as) 'a rigid policy of balancing the Budget each year, regardless of the state of trade'.[1] The task of the Chancellor is now far wider: 'to match our resources against our needs so that the main features of our economy may be worked out for the benefit of the community as a whole'.[2] Some would no doubt argue that, in theory at any rate, his responsibilities have enlarged still further since this principle was laid down in 1948. The growing appreciation of the need for public amenity and an environment appropriate to the twentieth century constitutes, fundamentally, a problem of public finance. So do the new and massive challenges in education, in housing and slum clearance, in the

[1] Cmd. 6527, 1944.
[2] Chancellor of the Exchequer, Budget Speech, April 6, 1948, *Hansard*, H. of C., Vol. 449, Col. 37. The new conception of the role of the budget as one of the chief regulators of the economy may be said to have originated in the war-time budget of 1941.

B

economics of old age, in the financing of the nationalized industries, in aid to the under-developed world, and in many other sectors of public and private expenditure. Taxation in its many varieties is now coming to be seen as one of the dominant political issues of the 1960's. Insofar as they prefer facts to political intuitions, Chancellors are going to depend more rather than less in future years on the statistics of income and wealth.

These are some of the reflections raised by a study of the sources of knowledge about the distribution of incomes. They are offered simply as evidence of the critical role now played by this body of facts. They lead us to ask: to what extent and in what respects do these statistics represent reality? How faithfully do they depict the changing constituents of income and wealth, and changes in rewards and ways of spending, giving and saving? How far do they take account of changes in the social and demographic structure of the population? How valid are the concepts and the data in relation to the uses to which they are put? These are not purely technical questions of statistical accuracy; they are fundamental questions about definition, measurement, analysis and interpretation.

These questions formed the starting point of this study which, in the main, is confined to the period since 1937–8. They remained the central *motif* throughout as the inquiry was pushed deeper; as the tools and the materials became more complex; and as account was increasingly taken of a multitude of social and demographic factors of change.

To appreciate what is involved in this limited approach it is first necessary to consider the works of others in the field of personal income distribution. In general, we need not recount the advances that have been made in the last twenty years in the measurement of the aggregate national income. These have been officially summarized in the Central Statistical Office publication *National Income Statistics: Sources and Methods* (1956).

In addition to the comparative analysis published by the Board of Inland Revenue in 1950, a number of studies have been made since the Second World War of changes in the distribution of incomes both before and after tax. These are referred to in the following chapter. We are principally concerned in this study with the pattern of distribution before tax. These are the basic data which fiscal policies must be related to and on which we rely for statements about equality and the capacity to bear taxation.

The two most recent and comprehensive inquiries which consider the distribution of incomes before tax are those by Mr Lydall (1955 and 1959) and Professor Paish (1957). We select these as the two leading studies in this particular field and we shall use them as models, so to speak, to illustrate some of the methodological issues involved. In later chapters we reproduce some of their findings; here, to avoid repetition, we need only refer briefly to their broad conclusions.

In general terms, both authors, working on the same basic material, were in agreement about the trend since 1938. They concluded that there had been a continuous movement towards greater equality of incomes before tax. Mr Lydall wrote of an 'exceptionally great' rate of change towards equality since 1938, little of which 'is attributable to the direct impact of taxation', and he argued that in the expansionist British economy of the late 1950's there was a permanent bias towards greater equality of incomes. After taking a longer view, and concluding that 'the trend in income distribution over the past two decades has been much more strongly egalitarian than in any previous period of our history', he proceeded to consider the specific causes of this trend and how far they were likely to operate in the future. Apart from the hazard of a major slump, he found no reasons, economic, social or political, to suggest that the movement towards greater equality, both before and after tax, would not continue as a permanent bias in an expanding British economy.[1]

Professor Paish, in his study, commented on 'this very remarkable redistribution of personal incomes before tax' between 1938 and 1955 and found, as Mr Lydall did, that the effects of tax increases had been relatively unimportant.[2]

Both authors made adjustments to their figures to take account of price changes but these did not upset their conclusions. Mr Lydall went further by making allowance for unallocated income and social service benefits, for indirect taxation, and for imputed income from undistributed profits. Again, these adjustments did not materially affect his conclusions. What dominates the results of these two studies is the continuous and rapid trend towards greater equality of incomes since 1938 and the relative unimportance to this trend of increases in direct Inland Revenue taxation.

No serious criticisms have since been made of these and other studies nor of the Board's own analysis, the conclusions of which have been accepted by the broad mass of opinion, specialist and lay, as confirming practically all the post-war writings on the subject of equality in Britain.

Sir Edward Boyle, Financial Secretary to the Treasury, expressed these views in the House of Commons in 1961 when he said 'we have a better and fairer distribution of incomes today than we had ten or eleven years ago'.[3]

[1] Lydall H. F., *Journal of the Royal Statistical Society*, 1959, Vol. 122, Part 1, pp. 33–5.
[2] Paish F. W., *Lloyds Bank Review*, 1957, *43*, pp. 11 and 14.
[3] *Hansard*, H. of C., February 23, 1961, Vol. 635, Col. 1027.

Confronted with such a remarkable measure of agreement among many economists and income statisticians there can be little doubt that those responsible for budgetary and fiscal policies in recent years have been profoundly influenced by these conclusions. They have provided the rationale for academic and political criticism of post-war social policies; they have furnished material for numerous public figures to plead for a more unequal society to stimulate economic behaviour and to promote a higher rate of growth; and they formed the justification for the budget of 1961 which raised the surtax limit to £5,000.

The purpose of this study is a severely limited one; to examine the statistical foundations on which law and opinion in the 1950's were based; to reinterpret the results of inquiries such as those made by Mr Lydall and Professor Paish; and to analyse a number of social and demographic factors of change in their bearing on these questions of equity and welfare. We begin by investigating some of the problems of definition.

CHAPTER 2

The Problem of Definition:
Income and Income Units

(I) THE DYNAMIC APPROACH

OVER the period covered by this study of broadly twenty years substantial changes took place in the social and demographic structure of the United Kingdom. For purposes of analysis in later chapters these changes are grouped, for convenience of arrangement rather than on any basis of principle, under three main heads. The first is concerned with demographic factors of change: the size of the population, and its age, sex and civil status structure. Some of the causes contributing to these structural changes and their consequences have to be considered in any analysis of income distribution. The most important ones are represented by the changes in the ratio of dependent non-earning units to earning units in the population; changes in the amount and age of marriage and the dissolution of marriages; and changes in rates of mortality and in the expectation of life of different groups in the population. All these factors which cause or express changes in the social and demographic structure of the national population must necessarily affect the income or tax population. The next chapter analyses these effects, and discusses their relevance to the statistics published by the Board of Inland Revenue and the interpretations drawn from these statistics by students of the distribution of income and wealth.

The second main group of change factors are largely social in character though they also contain a substantial component of demographic variables. Broadly this group includes: changes in the social and legal position of children, their age of entry into work, their possession of separate incomes and their ownership of capital; changes in the paid employment of married women by age, social class and income group; changes in the amount and distribution of unemployment by sex, age and occupational group; and changes in the age of retirement from work by sex, civil status, family structure, social class and income group.

Since 1938, the starting point of this study, far-reaching changes have occurred in the United Kingdom in all these spheres. The more important effects of some of them on the statistics of income distribution are also considered in later chapters.

Lastly, for our third heading, we examine a complex group of factors which are brought together under chapters 5 to 8. One of their common characteristics resides in the fact that they all involve some element of legal arrangement or manipulation of the constituents and sources of spending power—the ownership of wealth and the appreciation of capital as well as current income in cash and in kind. The importance of this part of the study is that all our knowledge about income-wealth in British society today is derived from two primary and distinct statistical boxes, one labelled 'income', the other 'wealth'. Moreover, what each box contains is information about individuals and not about families or wider kin groups. Yet, for an understanding of the changing distribution of income and wealth, classification and analysis by family and kin may be more significant in many respects than data presented in terms of the statistical and economic concept of the lone income unit the 'individual'.

Any system of direct taxation that takes account of personal circumstances must decide, however, what is to be the unit of taxation. That, in personal taxation, the unit will in some form be the individual is clear: but is it to be every individual separately, or is the married couple or the family the more appropriate unit? This question was raised in the Second Report of the Royal Commission on Taxation.[1] It went on to say that in the United Kingdom 'a husband and wife living together have always been treated as one unit, though in recent years this principle has been much modified in its application to a wife's earnings'.

In terms of fiscal policy, the Commission thought it right that the income of husbands and wives should continue to be aggregated.[2] The fundamental reason, though not made explicit, was that the family constitutes a common spending unit. Inexplicably, however, and despite the Commission's great concern about the family circumstances of taxpayers expressed in its recommendations, it did not propose that the incomes of children should be aggregated with those of their parents. Four members of the Commission did, however, recommend in a separate Reservation that the principle of aggrega-

[1] Cmd. 9105, 1954, p. 35.

[2] Although many wives do not know their husbands' incomes and husbands do not know their wives' incomes (Royal Commission, *Final Report*, Cmd. 9474, 1955, p. 323). In an Historical Note on the assessment of Income Tax on married persons prepared for the Royal Commission on the Income Tax in 1920 it was stated that after the supertax was imposed 'it was soon found in practice . . . that in many cases the husband did not know, and had no means of ascertaining, the income of his wife'. (*Minutes of Evidence*, App. 7 (d), Cmd. 615, 1920.)

tion should be adopted for all children under the age of sixteen.[1]

The historical development in fiscal policy of the concept of the individual, the family and the kinship group is a fascinating subject in itself. Here we are only concerned with the practice of the Board of Inland Revenue in its analysis and presentation of the statistics of income. How the data are presented is the question which immediately concerns us; the form such presentation takes does not of necessity have to follow the administrative and legal arrangements for the charging of taxation. The definition, employed by the Board for purposes of presentation, of the income unit in a society in which the structure and functioning of kinship relationships, legal, social and economic, differ over time is thus one of the central problems to be faced in studying the distribution of income and wealth. Secular differences in structure and relationships in the real world may invalidate comparisons of income distribution based on the concept of a single taxable unit. For this reason and also because of changes brought about by the use of various legal instruments, it becomes necessary in later chapters to examine the ways in which income and wealth may be split, rearranged and reallocated over time between family members and relatives by means of gifts, 'one-man company' incorporation, transfers, family settlements, discretionary trusts and other manipulative acts.

Allied to these problems of the splitting and spreading of income-wealth over time is the further theoretical and practical problem of distinguishing between the two boxes: what is income and what is wealth? This presents both social and legal questions of definition in terms of what goes into these boxes and what comes out in the form of statistical averages. For taxation and estate duty avoidance purposes, and for other social and legal reasons, income may be converted into capital, capital into income, and both into benefits that are neither money nor convertible into money. In each type of conversion, short-run, long-run or reversible, the time factor also enters. The redistribution of income-wealth over time, through 'spreading' over the life of an individual or of the family or of several kin generations, must clearly affect the shape and pattern of any income distribution which purports to show the facts for a single tax year. Whatever meaning the calendar year once had for economic men it has much less significance today. The consequences of more and more people anticipating the future were stressed by Keynes in his economic theory of 1936. The conception of a 'changing future influencing the present'[2] in equilibrium analysis is no less relevant to the

[1] Reservation by Messrs Bullock, Kaldor and Woodcock and Mrs Anstey, Cmd. 9105, 1954, p. 77.

[2] Keynes J. Maynard, *General Theory of Employment, Interest, and Money*, 1936, chapter 5.

study of the distribution of income and wealth. From the anticipation of a changing future to the manipulation of present and future is but a short and manageable step for those who have the expertise to follow the track. It is taken more readily when there is more certitude about the expectation of life for oneself and the members of one's family. Changing rates of mortality since 1938 brought about by the revolution of science in medicine thus represent one of the complex of factors which affect the pattern of income distribution.

The unit of time, the unit of income, and the definition of income itself are three inter-related issues which dominate this study. The manipulation and manoeuvrability of all three, independently and in combination, are examined in some detail in chapters 4 to 7. Chapter 8 is concerned with income in kind; it directs attention to some of the reasons why, as a result of the growth of 'fringe benefits', the Blue Book figures of salaries and wages are becoming increasingly inappropriate as a guide to income differentials. The effect of these analyses is to import a series of dynamic considerations into what is often statistically represented as a stationary world peopled by static income units.

These preliminary explanations concerning the role of social, legal and demographic factors in changing the real as well as the statistical pattern of income distribution since 1938 are amplified later. They are introduced here, though with some risk of repetition in mind, to provide a wider context in which to set an account of the materials and methods commonly employed in income distribution studies.

(II) THE BOARD OF INLAND REVENUE'S STATISTICS

Under the Inland Revenue Regulation Act, 1890, the Board is responsible for reporting to Parliament on 'the general care and management of the Inland Revenue duties'. This it does each year in an annual report presented by the Financial Secretary to the Treasury. It is this document which provides, and has always provided, the fundamental data on which all studies of the distribution of personal incomes and wealth in the United Kingdom have been based.

In its latest report at the time of writing (*BIR 103* for the year ended March 31, 1960) an account is given of the administration of the principal Inland Revenue duties in force, namely, income tax, surtax, profits tax, estate duty and stamp duties.[1] We are only concerned with the statistics for the first three insofar as they relate to *personal incomes before tax*. Only indirectly and incidentally are we interested in the tables on the classification of incomes after tax and

---

[1] A brief summary of the five income tax schedules is given in Appendix G for those who are unfamiliar with income tax law.

those which analyse the capital value of estates and estate duty receipts.

Though many of the statistics presented annually naturally relate to the exercise of administrative responsibilities, the Board has developed, over the last twenty years or so, a concern for the publication of data which analyse and classify the distribution of personal incomes. This function, which extends the Board's interests beyond its purely administrative duties, has led it to undertake from time to time a series of sample surveys and to analyse the results (as it does with its annual data) according to family circumstances and other characteristics.

These results are no doubt helpful to the Chancellor of the Exchequer for policy-making purposes. They allow, for instance, firmer estimates to be made of the cost to the revenue of contemplated changes in the level of taxation and personal allowances. The fact, however, that this basic material is published (and not retained simply as confidential data for the Chancellor) is one indication of the wider interpretation now placed by the Board on the manner in which it should discharge its responsibilities to the public at large as well as to Parliament. Another manifestation of its concern for extending its research and public information functions is the analysis it made, in its 1949 Report, of changes in the distribution of incomes between 1938–9 and 1948–9.[1] From this analysis, illustrated by two striking charts, the Board claimed that there had been 'a very considerable redistribution in incomes since pre-war' and that this change was 'most marked in the case of net incomes after tax'.

Although we shall have occasion later to criticize on many counts the Board's statistics it is proper to acknowledge here the contribution that its staff have made to the analysis, classification and publication of personal income data. In part, this material makes up for the lack of raw and unclassified statistics similar to those published by the Board in earlier periods. A study of the Board's reports for each year since 1908 shows that, before the First World War, there was a more liberal publication of primary—though often unclassified—data. Detailed statistics were presented annually, for example, on life assurance deductions, and the incomes of Government officials and other categories of employees.[2] This practice of releasing for public use large collections of statistical data was in line with the custom of Royal Commissions at that time in making available, to the benefit of later students of society, extensive appendices irrespective of the immediate topical relevance of the published material. Like other customs it is one which is sadly out of favour these days.

During the 1920's and the 1930's, the Board's statistics tended to

[1] *BIR 92*, p. 82.
[2] See, for example, *BIR 51–5* for 1908–12.

diminish in quantity, although it is quite remarkable, having regard to the increasing scope and complexity of direct taxation, how the main tables have retained for many decades the same form, and the statistics themselves the same definitions and treatment. The most important loss after the First World War were the statistics relating to the distribution of incomes by size. These were dropped for nearly twenty years because, in the words of the Board's Report for 1921, 'they could only be procured by incurring additional expenditure upon staff and machinery'.[1] During this period, when major political decisions were made on social and economic issues facing the country, there was little or no current information about trends in the distribution of personal incomes. The direction in which British society was moving in terms of income inequalities was not known. Ironically, it was when the country was most hard-pressed financially in the Second World War that staff and machinery were found. A special investigation of the returns of income furnished by taxpayers in connection with the assessments for 1937–8 was undertaken by the Board and the results were published in 1946.[2]

Apart from its yearly statistics which, as we shall see later, are unsatisfactory from the point of view of income distribution analysis, the first special surveys were undertaken by the Board in 1918–19 and 1919–20 in connection with the work of the Royal Commission on the Income Tax and the Committee on National Debt.[3] Administrative economies ruled out any further inquiries until the 1937–8 assessments were analysed. As four-fifths of these refer to income received in the previous year, the results purport to show, therefore, the distribution of incomes by size in the United Kingdom in 1936–7. To avoid ambiguities, however, and to follow the Board's practice, this inquiry will be referred to as the 1937–8 survey.

The survey was limited to total (undefined) incomes of £200 a year and over. An account of what must have been an immensely complex operation involving—if it was a complete census which is not clear from the report—over 3,000,000 'cases' takes up twenty-four lines in the Board's 83rd Report.

In 1950 the Board described this inquiry as an 'Income Census' and

[1] *BIR 64*, 1921, p. 79. Another serious casualty, ten years later, was the omission of published statistics of the amounts of dutiable settled property. These data derived from estates paying duty had been published by the Board for each year since 1897–8. Nothing has appeared since 1930–1. The importance of this information to our knowledge about trends in the distribution of wealth was explained by Mr J. R. S. Revell in an article in the *British Tax Review*, May–June 1961, p. 177.

[2] *BIR 83*, p. 28.

[3] *Report of the Royal Commission on the Income Tax, 1920*, Cmd. 615. Estimates in this Report were revised and subsequently published in *BIR 63*. Similar estimates for 1919–20 were published in the *Report of the Committee on National Debt and Taxation*, Cmd. 2800, 1927.

pointed out that as it was 'based primarily on the Income Tax return forms' it suffered from certain defects.[1] Although it reflected faithfully any errors or omissions in such returns nevertheless, said the Board, 'it provided a basis for subsequent work on the distribution of personal incomes, and the distributions produced in the Department since that date have been based on the 1937 Census brought up to date by reference to the annual statistics of assessments, etc.'.

One of the difficulties involved in furnishing statistics of income distribution is (to quote from the Board's Report for 1949) 'because Income Tax assessments are not made on an individual's total income but on each source of income separately. This difference is of no importance where there is only one source or where the subsidiary sources can be covered by personal allowances. But many people do have more than one source of income, and commonly more than one assessment is made on a particular individual. It is from these assessments (or in the case of PAYE the substituted office documents) that most of our statistical material is obtained. The task of deriving a distribution of total incomes from these figures is one of some difficulty attended by a risk of error.'[2] This risk is, of course, increased in the case of earning wives and other members of the family with individual and different sources of income.

In the brief report on the 1937–8 census no information was given on the extent to which these problems had been overcome or avoided. No attempt was apparently made to cross-check the results with other data on population, employment and so forth. For example, the fact that only 214,000 earning wives were reported was not evaluated in relation to other and independent sources of information.

Apart from an American economist, Professor A. M. Cartter, who was moved to remark that the constitution of the Board's figures 'remains something of a mystery'[3] they have not been subjected by income statisticians to any fundamental critical appraisal. Yet the 1937–8 census was—and still is—the fount of all our knowledge on the distribution of personal incomes in Britain before the Second World War. It was described by Mr Lydall as the 'first satisfactory estimate of the size distribution of incomes in this country'.[4]

For 1937–8, as a result of the census, a size distribution of total personal incomes was obtained. For the years between 1937–8 and 1949–50, only the amounts of income assessed under each schedule were known, and the size distribution of total incomes was estimated

[1] *BIR 92*, p. 82.
[2] *Ibid.*
[3] Cartter A. M., *The Redistribution of Income in Post-war Britain*, 1955, p. 137.
[4] Lydall H. F., *op. cit.*, p. 2.

by combining income under the different schedules in roughly the same proportions as in 1937–8. Throughout this period the Board assumed, despite one of the greatest upheavals in Britain's economic and social history, that this method of aggregating income from different sources held good. It not only made this assumption—or rather a complex series of interlocking assumptions—but it used the results to reach the widely publicized conclusion that over the period there had been 'a very considerable redistribution in incomes'.[1] No economist or income statistician in the United Kingdom has subsequently questioned this conclusion.

What was described as another 'special Income Census' was taken in respect of the year 1949–50. The first results were published in 1952.[2] Although given the same name, it was carried out quite differently from the 1937–8 inquiry inasmuch as it was based on a selected 10 per cent sample of all taxpayers (the income limit being £135). All their sources of income were said to have been aggregated 'by reference to their returns and other income tax documents'. The results from this sample were then grossed up to represent all those above the exemption limit.

The same procedure was followed in a third inquiry taken in 1954–5 the first results of which were published in 1957.[3] Here, however, the word 'Census' was dropped, and the sample was reduced to 5 per cent of taxpayers above the exemption limit of £155 in 1954–5. In both this and the previous inquiry the figures mostly referred to income received in the year of assessment. Only profits and professional earnings data related to the preceding year.

For both these surveys the Board said that (unlike the 1937–8 Report) the results had been 'compared wherever possible with independent estimates of the various totals'.[4] This check threw up a number of discrepancies, the most important relating to considerable deficiencies in the reporting of income from interest and dividends taxed at source; the omission of large numbers of married women in employment; and the omission of large numbers of children. We discuss later the significance of these particular discrepancies; meanwhile, it should be noted that the Board made no mention of them in connection with the 1937–8 inquiry. It is difficult, however, to know how much weight to give to these and other discrepancies because so little is known about the precise methods adopted by the Board in carrying out these surveys and in analysing the results. They have been reported with the utmost brevity; the 1937–8 inquiry was textually discussed in twenty-four lines; the 1949–50 census was dealt

---

[1] *BIR 92*, p. 86.
[2] *BIR 94*, p. 95.
[3] *BIR 99*, p. 66.
[4] *BIR 94*, p. 96 and *99*, p. 87.

with in under a page and a half; the 1954–5 survey in about two pages. In total, less than five pages have been devoted in over twenty years to describing the methods employed in funding for public use the basic data from which all our knowledge derives concerning the distribution of incomes by size.

Yet these three surveys represent the only serious attempts in over forty years to correct for some of the gross defects in the annual administrative statistics. As 'censuses of income', they may be regarded as at least of corresponding importance to the general well-being of the population as the censuses of population carried out by the Registrars-General. Nevertheless, while the definitions, concepts, methods, forms of tabulation and analyses used by the latter may be checked, compared from census to census, and discussed, the Board has published no information, for example, on sampling method, survey design, the problems of definition, the difficulties of linking or marrying schedule returns and much else besides. The technical problems of, for instance, drawing samples on a local basis must be extremely formidable in character, particularly as the possibility of error greatly increases at the top end of the income scale—the end which profoundly affects the shape of the whole distribution.

Later in this study it will be necessary to refer again to these gaps in our knowledge about the collection and analysis of the basic material. They are raised particularly in considering various problems of definition and terminology the importance of which, in the administration of tax law, was heavily stressed by the Income Tax Codification Committee in 1936. The fact that the word 'assessment' was then being used in eight different senses in the administration of the Statutes was one example used by the Committee to illustrate the anomalies that had evolved in Departmental practice.[1]

In discussing these problems of definition use will also be made of the tables and explanatory notes published by the Central Statistical Office in the National Income Blue Books. These, however, are secondary sources, as the data are based on the Board's statistics, although adjustments and extrapolations are made to take account of incomes falling below the exemption limit and other factors.

In an attempt to fill some of the gaps in knowledge—particularly those raised by the special income surveys—a questionnaire was drawn up and submitted to the Board in March 1961 (see Appendix A). It is admittedly a formidable document, running to seventy-eight items, and might well constitute a five-year programme of intensive research. It was thought wise, however, to make it as comprehensive as possible, partly to illustrate how little we know about the basic sources of information on the distribution of incomes and the effects

[1] *Report*, Vol. 1, Cmd. 5131, 1936, p. 13.

of age, sex, family settlements, discretionary trusts, covenants, pension schemes and many other factors. The questionnaire was submitted in the realization that the Board would probably be able to answer only a fraction of the questions. This has been the case. In the small number of instances in which additional information has been received, and which is gratefully acknowledged here, the replies have been incorporated in the text of this study. In general, the Board was not in a position to furnish additional statistical data. In some instances, the information provided served to correct a wrong interpretation by the author of the statutes or of the published material.

Finally, to complete this brief report on the Board's statistics on the classification of personal incomes, it should be noted that further tables have been published for the years 1955–6, 1956–7, 1957–8 and 1958–9. These are all estimates and are based on the statistics of income charged under the various schedules,[1] and on the Sample Survey of incomes from all sources in 1954–5. The pattern of distribution by income range for *all sources of income*, found in the 1954–5 Sample Survey, is assumed to apply without adjustment to all personal incomes in the years 1955–9. The implications of this assumption, which are not discussed in detail in any of the Board's reports, are so far-reaching that they probably invalidate in all but the crudest senses the results obtained.

If it is now the practice of the Board to hold a 'sample census' every five years the next one will relate to 1959–60. The complete results of such a census are unlikely to be available for public study until about February 1963. Until then, the projections of 1954–5 hold the field.

As mentioned in the preceding chapter, the two most recent studies of the distribution of incomes before tax are those by Mr Lydall (1955 and 1959) and Professor Paish (1957). Both are based on the Board's statistics and particularly on its two postwar surveys which, according to Mr Lydall, 'give a very full picture of the distribution of income by size'.[2] Other studies of various types since the Second World War have been made by Dr Barna (1945), Professor Cartter (1955), Professor Allen (1953 and 1957), Mr Seers (1951 and 1956) and Dr Rhodes (1951).[3] The work by Dr Barna explores the

---

[1] For Schedule E income the figures are estimated partly from a $2\frac{1}{2}$ per cent sample of tax deduction cards and partly from assessments, which are made in all cases where income exceeds £2,000 a year. Here again, no information is given concerning sampling methods and the linking of the results to assessments under other schedules and the 1954–5 projected distributions.

[2] Lydall H. F., *op. cit.*, p. 3.

[3] Reference should also be made to the Reservation by Messrs Woodcock, Bullock and Kaldor and Mrs Anstey in the *Second Report of the Royal Commission on Taxation*, Cmd. 9105, 1954.

pre-war statistics in considerable detail, and that by Mr Lydall provides a comprehensive introduction to the material on income distribution studies and methods of analysis.[1] The Central Office of Information Reference Booklet *The British System of Taxation* serves as a useful guide to the present-day system.

(III) THE DEFINITION OF INCOME

Later in this study we shall have much to say about the fundamental problem of defining income for the specific purposes of measuring the distribution of personal incomes by size. Clearly, this problem is largely governed by what the statutes define (or do not define) as income in cash or in kind for the purposes of taxation; how the Board uses its discretionary powers in the assessment and interpretation of what is personal income; and what decisions are reached by the courts concerning the borderline between income and capital.[2]

Income, like power, class and democracy, is susceptible of many definitions—or none at all. Yet, as Mr Kaldor has pointed out, it 'is not generally subjected to any searching or systematic analysis in economic textbooks'.[3] Mr Kaldor, who takes nothing for granted unlike most writers on public finance,[4] then proceeds to analyse, in a separate appendix to his book, 'the concept of income in economic theory'. In this discussion we rely heavily and gratefully on his analysis.

In the application of theory to statistical data on incomes there is, of course, no correct definition for all purposes. A choice has to be made, according to the purpose in hand and the availability of data, between a number of alternative definitions. At one extreme it is possible to define personal income as only that income, received in spendable form in a given year, which is actually spent on current consumption in the same year. Such a definition would exclude, for instance, all forms of deferred spending or saving from many categories of instalment contracts to life assurance premiums. The result of applying such a definition would tell us something about patterns

[1] The main sources of information about the distribution of incomes below the exemption limit are the publications of the Central Statistical Office, Lydall H. F., *British Incomes and Savings*, 1955, and Erritt M. J. and Nicholson J. L., 'The 1955 Savings Survey' in *Oxford Institute of Statistics Bulletin*, 1958, Vol. 20, pp. 113–152.

[2] 'There have been many cases,' said Lord Greene, 'which fall on the borderline. Indeed in many cases it is almost true to say that the spin of a coin would decide the matter almost as satisfactorily as an attempt to find reasons.' (Lord Greene in *British Salmson Aero Engines Ltd. v. IRC* (1938), 2 KB 482 at p. 498.)

[3] Kaldor N., *An Expenditure Tax*, 1955, p. 54.

[4] From this criticism we must except Mr Lydall and Mr Prest. The latter discusses, in his book *Public Finance*, the problem of distinguishing between 'true' assessable income and actual assessable income (1960, p. 257).

of current expenditure, but what we should learn about the distribution of incomes and wealth would be inadequate.

At the other extreme it is possible to conceive of a definition of income which would take account of all forms of income, personal to the individual, in kind as well as in cash; in some measure of all forms of saving from undistributed corporate profits to expected tax-free retirement lump sums and capital gains; all forms of re-allocated or 'split' income, present and future, to other members of the family or kinship, born and unborn. One obvious objection is that such a definition of income begins to assume the properties of a definition of wealth (or accretions of wealth) applicable, not simply to an individual or an 'income unit', but to a family or kinship group embracing perhaps three generations. But that is the crux of the matter today. As more income passes or is transmuted into forms of wealth or capital on a kinship basis the conventional income statistics become less and less meaningful in terms of the notions commonly attributed to them.

Only in a special sense was the Income Tax Codification Committee of 1936 concerned with this problem. The Committee considered a suggestion that its draft Bill 'should contain an express statement that income for the purposes of the Bill does not include any increase or increment of capital, on the ground that there is implicit in the existing law a general principle, which has been acted upon in many decided cases, that receipts of a capital nature are not liable to income tax. We decided not to include such a provision. "Income tax," as Lord Macnaghten said in his classic speech in *Attorney-General* v. *London County Council*,. "is a tax on income. It is not meant to be a tax on anything else." But what is capital is not more easy to define than what is income, and the futility of defining *ignotum per ignotius*, as well as a consideration of the complicated distinctions drawn in judgments of the Courts between fixed capital and floating capital, warned us that we should be on dangerous ground if we attempted to deal with the matter.'[2]

The majority of the members of the Royal Commission on Taxation (1952–5) did not venture far into this dangerous territory. They noted that the income tax code, unlike that of some other countries, contained no general definition of income, and they recognized that the major difficulties arose over that range of 'income' which does not admit of a clear dividing line between capital and income.[3] While apparently preferring to leave the concept vague and undefined for the particular purposes of their Report, the majority were, however, uneasily forced to distinguish between 'saving in general and saving

[1] (1901) AC 26 at p. 35; 4 Tax Cases 265, at p. 293.
[2] *Report*, Vol. 1, Cmd. 5131, 1936, pp. 23–4.
[3] Cmd. 9474, 1955, p. 7.

that is ascertainably related to the object of obtaining for the saver
and his dependents a reasonable provision in the case of retirement,
death or emergency'.[1] 'Dependents' were not defined; nor was
'emergency'. Yet both offer, to the cautious and well-advised man, a
wide range of choices and opportunities for 'saving'. But apart from
these reasons for exempting income from taxation it was considered
that 'a man ought to save . . . as a member of a society whose future
is pledged to economic progress'.[2] These views about the collective
duties of all men in modern society were critical in deciding the attitude of the Commission towards preferential tax relief for certain
forms of 'saving' and 'capital gains'.

'. . . no mere improvement of a person's financial or material
position is recognized as constituting income' was the general view
adopted by the majority of the Commission.[3] They thought, therefore, that no income should be recognized as arising unless an actual
receipt had taken place 'although a receipt may take the form of a
benefit having money's worth in kind as well as of money or of a
payment made to a third party in discharge of another's legal debt'.

These quotations from the Commission's *Final Report* illustrate
in a broad sense the practical problems of defining income for the
purposes of the Board of Inland Revenue's statistics on the distribution of income. Without this background, it is not possible to
grasp fully the significance of the criticisms we make later concerning
the Board's statistics and the studies by Mr Lydall, Professor Paish
and others. The problems may be summed up in the words of the
more incisive definition formulated in the Memorandum of Dissent
by Messrs Kaldor, Woodcock and Bullock:

In our view the taxable capacity of an individual consists in his
power to satisfy his own material needs, i.e. to attain a particular
living standard. We know of no alternative definition that is
capable of satisfying society's prevailing sense of fairness and
equity. Thus the ruling test to be applied in deciding whether any
particular receipt should or should not be reckoned as taxable
income is whether it contributes or not, or how far it contributes,
to an individual's 'spending power' during a period. When set
beside this standard, most of the principles that have been applied,
at one time or another, to determine whether particular types of
receipt constitute income (whether the receipts are regularly
recurrent or casual, or whether they proceed from a separate and
identifiable source, or whether they are payments for services
rendered, or whether they constitute profit 'on sound accountancy
principles', or whether, in the words of the Majority,[4] they fall

[1] Cmd. 9474, 1955, p. 20.
[2] *Ibid.*, p. 19.
[3] *Ibid.*, p. 8.
[4] *Ibid.*, p. 8.

C

'within the limited class of receipts that are identified as income by their own nature') appear to us to be irrelevant. In fact no concept of income can be really equitable that stops short of the comprehensive definition which embraces all receipts which increase an individual's command over the use of society's scarce resources—in other words, his 'net accretion of economic power between two points of time'.

Definitions of income giving expression to this basic principle have been offered by various writers on public finance, all of which, subject to minor differences, agree in regarding 'income' as the sum of two separate elements, namely personal consumption and net capital accumulation. In the words of one writer income can be looked upon either as '(a) the amount by which the value of a person's store of property rights would have increased, as between the beginning and the end of the period, if he had consumed (destroyed) nothing; or (b) the value of rights which he might have exercised in consumption without altering the value of his store of rights'. Hence income is the 'algebraic sum of (1) the market value of rights exercised in consumption, and (2) the change in the value of the store of property rights between the beginning and the end of the period in question'.[1]

The above definition focuses attention on two fundamental aspects of the concept of income which reflects the increment of 'spending power' or 'economic power' in a period. One is that income is a measure of the increase in the individual's command over resources in a period, irrespective of how much or how little of that command he actually exercises in consumption. The private choice of an individual as to how much he spends and how much he saves is irrelevant to this notion: income is the sum of consumption and net saving. The second is that 'net saving' (and hence income) includes the whole of the change in the value of man's store of property rights between two points of time, irrespective of whether the change has been brought about by the current addition to property which is saving in the narrower sense, or whether it has been caused by accretions to the value of property. From the point of view of an individual's command over resources, it is the change in the real value of his property which alone matters, and not the process by which that change was brought about.[2]

This statement sums up the essentials of the problem—described by Mr Kaldor in a picturesque phrase as the Iron Curtain between Capital and Income—which we shall meet again in many practical forms in later pages. How far does the existence of this Curtain, resting on the nineteenth century assumption that there is a firm boundary between capital and income, and continually changing in its texture and contours, affect the Board of Inland Revenue's

[1] Simons H., *Personal Income Taxation*, 1938, pp. 49–50.
[2] Cmd. 9474, 1955, pp. 355–6.

statistics of personal incomes and the conclusions that have been drawn from them?

Virtually all the writing on the distribution of incomes for lay and specialist audiences during the past twenty years has conveyed the general intent of the formulation by Messrs Kaldor, Woodcock and Bullock. This, as we shall see, differs in many fundamental respects from the present definition of income used by the Board in classifying and presenting the distribution of personal incomes, namely:

> Income before tax is all the income brought under the review of the Department, after certain deductions. It is after deducting losses and capital allowances in the case of profits and professional earnings and the allowance for repairs in the case of income from property; it is also after deducting National Insurance and superannuation contributions and other allowable expenses, mortgage interest and similar annual payments. It is before deducting the personal allowances or life assurance relief. It excludes income not subject to tax, such as interest on National Savings Certificates, National Assistance grants and certain National Insurance benefits and grants (unemployment, maternity, sickness, industrial injury, etc.).[1]

In the following chapters this definition, which determines the base of all our information about personal incomes, is analysed from the point of view of the studies that have been made of the Board's statistics.

---

[1] *BIR 103*, p. 74. For the purposes of the 1937–8 income survey the Board employed the following and much shorter definition: 'The income represents the net income of the taxpayer as computed for Income Tax purposes after deducting any charges (e.g. loan interest or ground rent) on that income.' (*BIR 83*, p. 29.)

CHAPTER 3

Social Structure and the Distribution of Incomes

I

ANY attempt to compare the size distribution of incomes in the United Kingdom between two or more points in time raises a number of problems. These are more formidable than is commonly supposed by most students of the subject, and no amount of sophisticated analysis can make up for inadequacies in the basic data. Statistical finery may hide but never remove the physical warts.

Initially, this chapter is concerned with two of these problems; the population at risk and its composition. We defer for later consideration another series of methodological problems which relate to what is being measured—in this case income. What we must first inquire into is the numerical size of the population with incomes, and how to determine, for comparative purposes, its 'parts' or income ranges or groups. It is essential to know as precisely as possible the total populations at risk for the years we are studying if we are to identify and compare the values of, for example, the top 1 per cent or 10 per cent of incomes. In pursuing this question it will ease the difficulties of explanation if, at the same time, we use as 'models' the two most recent comparative studies of income distribution in the United Kingdom, namely, those by Mr Lydall (1959) and Professor Paish (1957). In presenting some of their data and conclusions we are then led to ask: what is the composition, for each period, of the comparative 'parts' or income ranges? Any substantial changes in social structure could materially affect the conclusions drawn from a study which disregards the main characteristics of the units in the population. The changes we have in mind for a comparative analysis which spans a period of ten to twenty years—a period long enough for substantial structural changes to take place—relate to age, sex, civil status and occupational status (as defined for fiscal purposes). We ask this question about composition because we want to know whether we are comparing like with like when we place side by side the income fortunes of two 'parts' of similar size.

II

There is a considerable literature on the methodological problems of summarizing and describing in mathematical terms income distribution as a whole.[1] We are not here concerned with these problems in anything more than general terms nor, at the other extreme, with measuring changes over time in the value position of individual income units.

Much also has been written in the realm of theoretical studies of aggregate income distribution since the classical works of Ricardo and Marx. We have no pretensions to contribute to these writings, though we believe that in the analysis offered in this book certain veins of thought are opened up which may have some bearing on recent work by various authorities in the field of aggregate demand and macrodistribution. Changes in social structure, in the social definition of income and wealth, in distribution within and between families as distinct from individual changes, and shifts in the allocation or redistribution of income-wealth over a longer time scale, may have some bearing today on our knowledge of the relative shares of the total product of an economy accruing to various economic classes. But this is to anticipate later chapters.

The initial concern of this particular section is with the practical problems of measurement—the method of percentiles and inter-percentile groups used by Mr Lydall, Professor Paish and other investigators of income distribution data. The use of percentiles is, to quote Mr Lydall, 'a convenient way of comparing changes in the level of incomes of persons at equivalent positions in the income distribution . . .'[2] A similar method is to estimate the proportion of total income received by different inter-percentile groups. Mr Lydall explains in his paper how he made such estimates and we quote from his account:

There are several convenient graphic methods of comparing income size distributions, of which the Pareto and Lorenz curves are the best known. Some writers also use the Gini coefficient to summarize the shape of the Lorenz curve. My intention, however, is not so much to summarize the distribution as a whole as to consider changes in the shape of the distribution in its different parts. For this purpose the Lorenz curve tends to be too insensitive an instrument and I have used instead a number of different curves drawn on double-logarithmic paper. With their assistance it is possible to establish fairly accurately the position of each percentile in the top half of the income distribution, and, in some of the

[1] For references see Aitchison J. and Brown J. A. C., *The Lognormal Distribution*, 1957.

[2] Lydall H. F., *op. cit.* (1959), p. 5. For explanation of percentiles, see note on p. 53.

post-war years with somewhat less accuracy, down as far as the
seventieth percentile from the top. In the same way we can measure
the proportion of total income received by each inter-percentile
group. Thus it is possible to follow the changes over the years in
the absolute and relative incomes of particular percentiles, and in
the proportions of total income received by particular percentage
groups of the population. The method used here is similar in many
aspects to that used by Paish (1957) in his stimulating paper.[1]

Applying his method to the basic Board of Inland Revenue data,
Mr Lydall then produced figures for the four years, 1938, 1949, 1954
and 1957. We give below in Table 1 his results for 1938 and 1957.

TABLE 1
PERCENTILES OF ALLOCATED INCOME BEFORE TAX

	1938	1957
Number of incomes of £50 a year or more	24,000,000	26,100,000
Percentile[a]	£	£
First	1,140	2,450
Fifth	393	1,180
Tenth	266	940
Twentieth	185	792
Fiftieth[b]	(110)	512

Notes: [a] Numbered from the highest downwards.
    [b] 'Less reliable estimates are put in brackets' (Lydall H. F., *op. cit.*
    (1959), p. 7).

The fiftieth percentile represents the median or central value of the
whole distribution. Since these figures were, as Mr Lydall points out,
'obtained by graphic interpolation they may deviate by 1 or 2 per
cent from their true values, but the broad pattern of the changes in
the relative position of the different percentiles would not be affected
by any inaccuracies of this sort'.[2]

The next table (Table 2) is extracted from the second of Mr
Lydall's key tables on income before tax:

TABLE 2
PERCENTAGE OF ALLOCATED INCOME BEFORE TAX
RECEIVED BY SPECIFIED INTER-PERCENTILE GROUPS

Inter-percentile Group	1938	1957
Top 1 per cent	16·2	8·0
Second to fifth per cent	12·8	10·2
Sixth to tenth per cent	9·0	9·8
Eleventh to twentieth per cent	12·0	13·5
Top 5 per cent	29·0	18·2
Top 10 per cent	38·0	28·0
Top 20 per cent	50·0	41·5

[1] Lydall H. F., *op. cit.* (1959), p. 5.
[2] *Ibid.*, pp. 6–7.

The top 1 per cent of incomes comprised 240,000 units in 1938 and 261,000 units in 1957.

Before commenting on these tables we give extracts for 1938 and 1955 from Professor Paish's key table:

TABLE 3

DISTRIBUTION OF PERSONAL INCOMES BEFORE TAX[1]

Income Range	1938 %	Range of Income covered £	1955 %	Range of Income covered £
1. 1st 100,000	11·7[2]	2,070 & over	5·3	3,850 & over
2. 2nd ,,	3·6	1,260–2,070	2·4	2,700–3,850
3. 3rd ,,	2·6	970–1,260	1·8	2,150–2,700
4. 4th ,,	2·0	795– 970	1·5	1,825–2,150
5. 5th ,,	1·6	685– 795	1·3	1,625–1,825
6. 1st 500,000	21·5	685 & over	12·3	1,625 & over
7. 2nd ,,	6·3	450– 685	5·1	1,175–1,625
8. 1st million	27·8	450 & over	17·4	1,175 & over
9. 2nd ,,	8·2	300– 450	7·8	910–1,175
10. 3rd ,,	6·0	238– 300	6·4	800– 910
11. 4th ,,	5·1	205– 238	5·7	730– 800
12. 5th ,,	4·5	184– 205	5·3	680– 730
13. 1st 5 millions	51·6	184 & over	42·6	680 & over
14. 2nd 5 ,,	16·8	123– 184	22·6	510– 680
15. 1st 10 millions	68·4	123 & over	65·2	510 & over
16. Remainder	31·6	Under 123	34·8	Under 510
Total Incomes	100·0		100·0	

It will be seen that, unlike Mr Lydall, Professor Paish uses income groups of constant size for the top ten million incomes; the 'remainder' forms one group of a different size in 1955 from that in 1938. We return later to this problem of the population and its breakdown into groups.

While Professor Paish uses the term 'personal incomes before tax' and Mr Lydall 'allocated income before tax' it is clear, from re-

[1] 'The table covers only attributable personal income.' (Paish F., *op. cit.*, pp. 7–9.)

[2] In re-calculating some of the figures in Table 3 to ascertain the total income base (no estimates are given in the published paper) a number of discrepancies were noted. For example: the 1938 percentage for the first 100,000 should be 11·4 not 11·7. According to table 31 of the 1958 *NIBB*, 11·7 represents the first 105,000 incomes. To obtain the correct percentage for 100,000 it is necessary to subtract 5,000 incomes from the number given (46,000) for the range £2,000–£3,000.

working some of their results, that they are both based on the same National Income Blue Book tables which, as we have already seen, are derived from the published Board of Inland Revenue data. Certain adjustments are made in the NIBB tables by the Central Statistical Office; these are discussed later.

The figures given in the above three tables are basically those which lead both authors to conclude that over the period of seventeen to nineteen years there has been a continuous trend towards greater equality of incomes before tax. Before examining this conclusion we must consider the problem of population size.

III

Whatever method is used to measure the distribution by 'parts'— percentiles, inter-percentile groups or constant size groups plus 'remainder'—it is fundamental to establish the total population at risk. *Who was present in 1938 and in 1955-7* is the question we must ask. Unless we know to a fair degree of accuracy the total number of incomes we cannot identify with any precision the percentiles and the size of the groups. Nor can we generalize about the 'remainder' unless we know the size of the universe. What is also essential to any discussion of the 'shape' of the distribution of incomes is to get the numbers straight for the two extremes—the thin top tail of 1 per cent to 5 per cent of incomes and the mass of low incomes. To fix these extremes for size and position we must know the total of incomes. When we have these facts, however, we cannot draw any conclusions before defining income and an income unit. Nevertheless, we have to take one step at a time and see what others have made of the population question. For the moment then our discussion is based on (but without necessarily accepting) the definitions employed by the Board of Inland Revenue and accepted by those who have used the Board's data.

For the post-war years Government statisticians have made estimates of the number of income units receiving an income of £50 or more. The total for 1957 is given as 25,900,000 (later adjusted to 26,100,000[1]). The Board of Inland Revenue's estimate of the 'number of individuals with total incomes above the exemption limit' (£180) is 20,900,000 for 1956-7.[2] Assuming for the present that the definitions used are identical (it should be noted that the Board uses the term 'individuals') it follows that there were 5,200,000 'individuals' or 'units' with incomes between £50 and £180. Mr Lydall states that the total of 26,100,000 (and other totals for 1949 and 1954) is 'broadly equivalent' to the number of married couples and of single

[1] *NIBB* for 1960, table 22.
[2] *BIR 101*, table 21.

persons aged 18 and over in the whole population.[1] He thus uses this method to arrive at the number of incomes.

While it may be of some value in relation to the composition of the population in certain years the use of the method for 1938 and for various years in the 1950's does, however, yield some puzzling results. The official population estimates for 1958 may be taken as an illustration.[2] If we add to the figures given in Appendix B for the number of married couples in 1958 (*a*) the number of single and widowed men and women aged 20 and over and (*b*) two-fifths of the number of single men and women aged 15–19 we arrive at a total of 24,747,450. It cannot be said that this is even 'broadly equivalent' to the Government statisticians' estimate of 26,100,000 income units. This being so, there is no reason to suppose that Mr Lydall's estimate of 24,000,000 income units for 1938 (also arrived at by taking the number of married couples and single persons aged 18 and over) is any nearer the mark.

It would in fact be no more than a passing coincidence if, in a particular year, the number of income units (as defined by the Central Statistical Office) was roughly equivalent to the number of married couples and single persons aged 18 and over. For this to happen, the number of potential income units with incomes under £50 for the year in question would have to be approximately equal to the number of persons under 18 with incomes over £50. Even supposing that this were true in 1957, it is unlikely that it would also hold good for 1938, in view of the changes in the proportions at work in different age groups, the levels of social security payments (especially pensions) and the widespread adoption between 1938 and 1957 of devices for transferring income from parents to children.

A particular drawback of the method of estimating the population at risk used by Mr Lydall is that it does not take account of changes in the size of the group of earners aged 15–17. In 1938 there were approximately 2,382,000 young people aged 15–17 of whom a very high proportion were in the labour force, employed and unemployed. In 1958 this group numbered only about 1,500,000 of whom a smaller proportion were in the labour force because of the fact that more of them were staying on longer at school.[3] Nevertheless, practically all the young people who were at work in 1958 must have been earning more than £50 a year and it is probable that their average income was in excess of that received by about one-third of

[1] Lydall H. F., *op. cit.* (1959), p. 6.

[2] The increase in the total population between mid-1957 (the last year in Mr Lydall's study) and mid-1958 was very small—approximately 225,000.

[3] The Government Actuary estimated that there were about 1,350,000 young people aged 15-17 paying National Insurance contributions in 1958. (*Report by the Government Actuary on the Second Quinquennial Review*, 1960.)

retirement pensioners.[1] In addition, there may well have been 100,000 or more university students, grammar, technical and public school children of 15–18 who earned more than £50 a year and who may not have been counted as employed persons. How all this income is treated by the Board in its tables it is quite impossible to tell from the annual reports. The question is, of course, related to the problem of definition; of claims for child allowances, and of 'missing' children —all matters which are discussed later.

As we have seen, Mr Lydall for the purposes of his study worked on the basis of 24,000,000 income units in 1938 with incomes of £50 and over.[2] Mr Seers, in his study of *The Levelling of Incomes since 1938*, estimated the figure at 23,500,000.[3] Professor Paish, who gave no information about method, estimated that there were 'probably well under' 25,000,000 incomes in 1938—but he included incomes below £50. This figure would mean, however, assuming as we must in this context that married couples were counted as units, that there were something like one million adults in the United Kingdom in 1938 without any income at all.[4] Such a conclusion is, of course, highly improbable.

These estimates and calculations, when examined and compared, are so puzzling and difficult to reconcile that it is necessary to analyse more closely the structure of the population of the United Kingdom in 1938. This is attempted in Appendix B which also includes comparative figures for 1958. The analysis is set out in terms of 'potential tax units' to accord with the Board's assumed definitions and 'total individuals' to achieve a reconciliation with the Registrar-General's statistics.

From the results of this analysis we can see something of the remarkable changes that took place between 1938 and 1958 in the demographic and social structure of the population. It is no exaggeration to say that these changes as a whole have been virtually ignored by all those who have studied the distribution of incomes since 1938.

If we are only concerned with estimating the two universes of 'incomes' (or 'potential tax units') the change has not been striking. Looked at from this point of view—the fiscal view of population change—the total of 'potential tax units' rose during the twenty years by only 303,000 or 1 per cent on the 1938 figure of 25,636,000.

---

[1] See *Annual Reports of the National Assistance Board.*
[2] Lydall H. F., *op. cit.* (1959), p. 6.
[3] Seers Dudley, *op. cit.*, p. 32.
[4] Paish F., *op. cit.*, p. 8. Professor Paish may have excluded income from unemployment benefit, public assistance and other grants but, even so, there would still remain a large number of incomeless adults. In Appendix B a total is reached of 25,636,000 'potential tax units' in 1938. This total excludes children and single people under age twenty with unearned incomes and separately assessed married women with earned and/or unearned incomes.

This does not suggest that any great error would be introduced by applying, for comparative purposes, the method of percentiles and inter-percentile groups as used by Mr Lydall, Professor Paish and other analysts of income distribution. The size of the parts and the position of the percentiles in 1938 and 1958 would not differ very much.

To accept such aggregate figures as these without further inquiry would, however, hide from view the very remarkable changes that are depicted in Appendix B and which, in a variety of complex ways, must radically alter the pattern of income distribution. We may summarize the main changes in a series of statements:

1. While the number of 'potential tax units' rose by 314,000 or 1 per cent the total population increased by 4,365,300 or 9 per cent and the number of married people by 4,781,000 or 22 per cent.

2. In addition to this increase of 4,781,000 in the married population, there were also increases of 818,000 or 27 per cent in the number of widows, widowers and divorced persons, and of 1,683,000 or 15 per cent in the size of the dependent age group of children and young persons not in the labour force.

3. Against these increases (totalling 7,282,000) there was a decline of 2,906,000 or 25 per cent in the number of single people aged 20 and over and aged 15–19 in the labour force. This decline included a remarkable fall in the number of single females aged 20 and over of 1,245,000 or 28 per cent, reflecting the great increase in marriages since 1938.

As important as the shifts in marital condition and 'dependency' in terms of taxable units were the changes that took place in the age and sex structure of the population between 1938 and 1958. A few of the more significant changes are singled out because of the far-reaching effects they are likely to have had on the pattern of income distribution:

4. The total population aged 65 and over increased by 1,881,000 or 46 per cent during the twenty years. This was made up of an increase in males of 611,000 or 35 per cent and an increase in females of 1,270,000 or 54 per cent.

5. Of special importance is the increase in the elderly female population because of the differential tax treatment involved in changes in marital condition, the effects of the inheritance of capital by widows, transfers among receivers of unearned incomes, capital gains and other factors. In the population aged 65 and over, the number of single females rose by 184,000 or 45 per cent, that of married females by 382,000 or 46 per cent, and that of the widowed and divorced by as much as 704,000 or 63 per cent.

Among all the social factors affecting the distribution of incomes one of the most profound in recent years has been the increase in the paid employment of married women. Little is known about the position in 1937–8 but it is probable that the number of married women of all ages at work lay between 500,000–800,000.[1] Including more than 138,000 self-employed, it was 'probably over 4,000,000' in 1954–5 counting as separate units those who were married, widowed or divorced during the year, as well as women married throughout the year who entered or left the labour force during the year.[2] To this figure should be added an unknown number of wives whose earnings or profits were below the tax deduction card limit of £165 a year. Of the total of over 4,000,000, something like 1,500,000 were 'missing' from the Board's classification of incomes by ranges because their earnings were omitted or understated in the husband's tax returns.[3] (This particular problem of 'missing wives' is further explored in a later chapter concerned with the definition of income units.)

This great increase between 1938 and 1958 of around 3,500,000 in the number of earning wives resulted not only from the fact that far more wives entered paid employment and received profits and professional earnings (some in the form of 'one-man companies') but also from the fact that there were far more married women 'at risk'. The figures for those age groups in which the proportion of married women at work is known to be relatively high[4] are shown below:

[1] According to the census of 1931, 10·4 per cent of married women were reported to be gainfully employed in England and Wales and 6.4 per cent in Scotland. By 1938 the number of married women had substantially increased (to a total of 8,324,000 at all ages under 55 in Great Britain) and the proportion gainfully employed may well have declined. According to estimates made by the Statistics Department of the Ministry of Labour in 1950 (personal letter November 25, 1950) there were at July 1937 872,200 married, widowed and divorced women (using 'Mrs' in front of their names) insured under the Unemployment Insurance Acts in Great Britain. A high proportion were probably widows as there were in 1938 2,227,000 widows at all ages over twenty in the UK. In 1939 the total of insured women aged 14–59 in Great Britain (industries and exemption limit as in 1945) was 4,325,000. This figure has to be seen in the context of a total of 8,003,000 single (excluding those at school and university), widowed and divorced women aged over fifteen in the UK. Mr Colin Clark, who did not distinguish the number widowed, put the number of employed married women at about 780,000 (excluding those unemployed) (Clark C., *National Income and Outlay*, 1937, p. 289). The number of earning wives in the UK in 1937–8 in all cases where total income exceeded £200 a year was reported to the Inland Revenue as 214,473 (*BIR 83*, p. 33).

[2] *BIR 101*, p. 71 and table 63.

[3] *BIR 100*, p. 80 and *101*, p. 70.

[4] An analysis of the statistics of married women in employment published in the *Ministry of Labour Gazette* (Aug. 1952) shows the following percentages for Great Britain: age 20–24, 40 per cent; 45–49, 30 per cent; 50–54, 25 per cent; 55–59, 18 per cent. By 1958 these proportions had risen to: 20–24, 44 per cent; 45–49, 35 per cent; 50–54, 32 per cent; 55–59, 26 per cent (estimated from National Insurance data. Stewart C. M., *British Journal of Sociology*, 1961, Vol. XII, No. 1, p. 5).

340,000 more married women aged 20–24 in the UK in 1958 than in
1938—61 per cent increase
648,000 more married women aged 45–54 in the UK in 1958 than in
1938—29 per cent increase
437,000 more married women aged 55–64 in the UK in 1958 than in
1938—27 per cent increase
Total: 1,425,000 more married women of these ages in the UK in 1958 than in
1938—32 per cent increase

IV

We have so far identified a number of major changes in the social
and demographic characteristics of the population and have given
evidence to show their importance in determining the shape of the
personal income distribution. It is now possible to return to the
questions posed earlier in this chapter. First we deal with the problem
of the size of the income population. As Mr Lydall says in his paper:
'In order to identify the percentiles it is, of course, necessary to know
the total number of incomes.'[1]

Working within the conventions which we have had to assume
have been consistently employed by the Board of Inland Revenue
since 1938, we have shown that changes in the potential tax unit
population are of quite a different order to changes in the total adult
population or in the number of income receivers and earners. To
accept the former (or some variant of it) as the 'income universe' can
yield results which do not reflect the real changes taking place in the
number of individual income-earning and receiving units.

It is also evident that the results obtained by other investigators
using different methods of estimating the total number of incomes
vary very considerably. Mr Lydall in accepting the CSO and BIR
estimate of total incomes of £50 a year or more for 1957 and relating
it to his estimate for 1938 (derived by adding together married couples
and single people aged 18 and over) arrives at an increase of 2,100,000.
But if the adjustments we made earlier are used instead the increase is
only 747,450. Professor Paish's study of all incomes indicates a rise
from 'well under' 25,000,000 in 1938 (implying the existence of
something like 1,000,000 adults without any incomes at all) to
26,200,000 in 1955. If 'well under' means, say, 24,750,000, this
implies a rise of 1,450,000. The estimates used by Mr Seers in his
study, also covering all incomes, give the total number of incomes as
23,500,000 in 1938 and precisely the same number in 1947.[2] The CSO
and BIR estimate of the number of incomes of £50 a year or more in
1949 was 26,100,000.[3] Nine years later it had risen by only 150,000

[1] Lydall H. F., op. cit. (1959), p. 6.
[2] Seers Dudley, op. cit., p. 34.
[3] NIBB, 1960, table 22. No estimates given for 1947–8.

to 26,250,000. Yet during this period the total population increased by 1,750,000, the average number of employees coming under 'Pay-as-you-earn' by 2,500,000,[1] and a very substantial part (perhaps one-half or more) of the remarkable changes in the social and demographic characteristics of the population illustrated in Appendix B had taken place.

Quite apart from all the ambiguities and contradictions which, as later pages show, surround the definitions of 'income', 'income units', 'children' and other 'dependents', it is difficult to have much confidence in these estimates of the income universe while there exist many unexplained discrepancies and such wide differences in the results. If the totals are wrong, then the positions assigned to the percentiles and the place and size of the inter-percentile groups must also be wrong. Such errors could have a greater effect in the determination of the size of the top and bottom incomes—the top tail and the bottom mass of low incomes in the pattern of distribution and the fundamental basis of all the statements about changes in the inequalities of income. A larger number of total units lowers, for example, the position in the income ranges of the 1st and 5th percentiles. Conversely, a smaller number of units raises the position of these percentiles.

We turn now to consider the differential effects of changes in the social and demographic characteristics of the population on different income ranges or 'parts' of the universe. The relevant and important tables to examine in this connection are Tables 1, 2 and 3 (given earlier in this chapter) taken for purposes of illustration from Mr Lydall's and Professor Paish's studies. The major conclusions drawn from these tables about 'strong egalitarian trends' are heavily influenced by the behaviour of the relatively small top units in the distribution. Thus, Mr Lydall shows that the share of allocated income before tax of the top 1 per cent fell 'catastrophically' (as he puts it) by half from 16·2 per cent to 8·0 per cent between 1938 and 1957.[2] This fall of 8·2 per cent accounted for nearly all the fall of 8·5 per cent of total allocated income in the share of the top 20 per cent of the income unit universe. The bottom 80 per cent thus gained an additional 8·5 per cent of the total allocated income.

Much the same pattern is exhibited in Professor Paish's table which is dominated by the change in the share received by the top 300,000 incomes (a fall of 8·4 per cent in a total fall of 10·4 per cent in the share of the top million). It should be noted that Professor Paish's top groups for his two years (1938 and 1955) contain the same number of tax units, while his bottom group 'remainder' embraces something like 1,450,000 more tax units in 1955 than in 1938. Unless they

[1] *BIR 102*, p. 67.
[2] Lydall H. F., *op. cit.* (1959), p. 14.

were all unemployed or retired and without any pensions, benefits or public assistance, it is difficult to believe that the addition of 1,450,000 tax units to a 'remainder' population of 14,750,000 in 1938 could not, by itself alone, increase the share of total income going to the 'bottom' groups and, consequently, reduce in proportion the share received by the top groups unaltered in size. Quite simply, an expansion of this order of magnitude in the size of the 'remainder' tax unit population must produce—other things being equal—a 'strongly egalitarian trend' in the pattern of distribution. No alternative statistical effect is logically possible.

In June 1938 there were 1,885,000 insured workers registered as unemployed. To this figure there has to be added some element for the 1,300,000 or so people on public assistance. Taking the year as a whole and making allowance for double-counting, there may well have been some 3–4,000,000 different individual tax units in a total population of 24–25,000,000 units who experienced some considerable spell of unemployment during the year. Most of these 3–4,000,000 units must have fallen in the 'remainder' or bottom group of the income universe. By contrast, in June 1957 there were fewer than 300,000 individuals registered as unemployed.

By giving work to the workless a statistical table may be produced which can be interpreted to show that the distribution of incomes has become markedly more egalitarian. A trend is apparently established which is then projected, fashion-wise, into the future. Perpetual repetition is thus attributed to historical experience. In some senses, of course, a society with full—or nearly full—employment is a more egalitarian society than one in which a fifth of its bottom mass of income-receivers are wholly or partially unemployed. But this is generally not the sense which readers of income distribution studies have in mind (or are specifically directed to bear in mind) when they consider the notions of more inequality or less inequality in British society in recent years.

While changes in the incidence of unemployment have materially influenced the fortunes of the bottom mass of low incomes since 1938, different factors have affected the composition and statistical behaviour of the top 1 per cent. In Mr Lydall's tables this group numbered 240,000 tax units in 1938. The corresponding figure for 1957 is 261,000. The allocated income before tax of the first percentile is given by Mr Lydall as £2,450 (1957) and £1,140 (1938). This means that in 1957 there were only 261,000 units with incomes of £2,450 and over.[1] What were the social and demographic characteristics of these units and in what respects did they differ from the 240,000 units in 1938? Such questions are, of course, applicable to other in-

[1] The *NIBB* for 1958 sets out the 1957 statistics (table 31). These were used by Mr Lydall. In both the 1959 and 1960 editions they were substantially revised:

come groups but in our analysis of the characteristics singled out in Appendix B we shall pay particular attention to their relevance to the top and bottom incomes.

We can attempt a rough systematization of these characteristics— or factors of change—as an indication of how they may influence in a variety of concordant and discordant ways the pattern of distribution. We do this under four headings:

Factors which tend to contract the tax universe
(*a*) More marriages (1 unit replacing 2)
(*b* More remarriages (1 unit replacing 2)
(*c*) A smaller number of the population in the labour force (0 unit replacing 1)

Factors which tend to expand the tax universe
(*d*) More children receiving unearned income (1 unit replacing 0)
(*e*) More divorces (2 units replacing 1)
(*f*) A higher number of the population in the labout force (1 unit replacing 0)

Factors which tend to 'bunch' units around the central value
(*g*) More marriages and fewer broken marriages among the lower and middle range income groups. This effect operates in the context of an increase in the employment of wives, and falling mortality rates (more income-receiving units joined together as one and remaining joined together into retirement).
(*h*) A higher number of retired units receiving pension increases for dependant wives (values of 1 increased)
(*i*) A smaller number unemployed in the labour force, both male and female (values of 1 increased)

Factors which tend to disperse units to one or both extremes of the distribution
(*j*) The reverse of factors (*g*) (*h*) and (*i*) (values of 1 decreased and more 2 units of lower value in place of 1)

	1958 Blue Book	1959 Blue Book	1960 Blue Book
£20,000 and over	3,000	3,000	3,000
£10,000–£20,000	12,000	12,000	12,000
£5,000–£10,000	51,000	52,000	54,000
£3,000– £5,000	114,000	120,000	126,000
£2,000– £3,000	198,000	218,000	205,000
	378,000	405,000	400,000

(*k*) A higher number of minors in the labour force or receiving unemployment assistance in their own right (more 1 units with low values)

(*l*) More marriages and remarriages in a given year among women with separate incomes (more lower value units in respect of past years). A similar effect would result from more deaths among husbands (more 2 units, 1 relatively low, 1 relatively high, replacing 1 higher unit in a given year).

The factors of demographic and social change grouped under these four categories are not, of course, exhaustive. More could be added. They are listed as illustrations of the wide variety of inter-connected factors which, operating in different directions and with different effects, can in combination profoundly change the pattern of income distribution as depicted in official statistics. With long and patient research and with far more basic data published by the Board of Inland Revenue it should be feasible to quantify the effects of at least some of these factors. But it is impossible to do this in the present state of the statistics. The formidable nature of the questionnaire submitted to the Board in March 1961 indicates the extent of our continued ignorance. All one can thus do is to speculate about the direction of certain major factors and to offer one or two general hypotheses.

When two income-receiving units are joined together in marriage or remarriage a single and higher income replaces (on the assumption of joint assessment) two lower ones. Conversely, death or divorce generally lowers the value of a unit or units (even among the pensioned population this effect operates). In short, the marriage and divorce habits and mortality experience by age, sex, marital condition and income range can influence to a far-reaching extent the shape as well as the size of the income distribution. As we have already said, the official statistics do not allow these influences to be detected for the universe as a whole. More important still, we can learn nothing about their differential effects at different levels of income. In comparisons over time, if there is, for example, a larger proportionate increase in the employment of working-class wives than middle and upper-class wives the income distribution table for the second period will look more egalitarian. All the evidence suggests that between 1938–58 this was one of the powerful factors which, along with the decline in unemployment among married and single men and single women, led to a statistical picture interpreted by Mr Lydall, Professor Paish, Mr Seers and others as 'strongly egalitarian' and as depicting an 'exceptionally great' diminution of inequality. While the employment of working-class wives has had in reality, of course, an equalizing effect, the statistical presentation of

D

the effect tends to exaggerate the amount of equalization. It does so because of the substitution of one higher value unit for two lower ones—a process accentuated since 1938 by the trend towards more and earlier marriages.

Some investigators of income distribution have projected their conclusions into the future by suggesting that in the economic expansion of 1938–57 there was, on the basis of evidence derived from the Board of Inland Revenue data, some sort of permanent and built-in bias towards greater equality of incomes. It should, however, be pointed out that there is now little scope for the further employment of wives and mothers (probably none at all if the trend of fertility in recent years continues its upward path); nor can we repeat all over again a reduction in unemployment of 2–3,000,000; nor the remarkable reductions in differential mortality under the age of 60 since 1938.

The phenomenal rise since 1938 in the amount of marriage and in the number of earlier marriages is one of the major factors in producing a statistical illusion of greater income equality. It is, nevertheless, a cause for surprise that the full significance of one of the most remarkable social changes of recent decades has been virtually ignored by practically all those who have commented on the income distribution statistics. Large numbers of girls and young women, formerly appearing as low value income units, disappeared completely from the statistics. Many reappeared later to inflate for a time the value of the units for married couples. But here again there are limits—which we are now rapidly approaching—to any further increase in the amount of marriage over the age of 21. Even the trend from single residential domestics in the 1930's to married non-residential charwomen in the 1950's can help in this process of statistical deception. But it cannot continue to help indefinitely.

We may now consider some of the possible effects of these factors specifically in relation to the top 1 to 5 per cent of incomes in 1938 and 1955–8. It is likely that these groups contained in the latter years, absolutely and relatively:

(a) More single income units represented by more elderly and wealthy widows due to the lengthening expectation of life among Social Class I women (of all estates exceeding £50,000 in net capital value in Great Britain in 1958–9 33 per cent were left by women).[1] These units, though high in value, would not be as high as the units which would have been included if the husbands had survived and

---

[1] *BIR 102*, tables 106–8. It is also interesting to note that an analysis in 1959 of Midland Bank shareholders showed that 55 per cent were women and 32 per cent men, the remainder being joint accounts and institutions (Cummings G., *The Times*, August 4, 1959). A similar study of stockholders by sex of the British-American Tobacco Company also reported a higher proportion of women (Cummings G., *The Times*, September 29, 1960).

had not split both capital and income before death among their families.

(*b*) More single income units represented by more divorced men and women in Social Class I relative to divorce rates in Social Classes IV–V.

(*c*) Proportionately fewer employed married women relative to the increase in the employment of wives of men in income ranges below £1,000.[1]

(*d*) More wives with 'professional earnings and profits' (and possibly separately assessed[2]) relative to 1938 and relative to lower income ranges. The sample census of 1954–5 threw up some remarkable figures in this connection. Of 43,500 married couple units in the income range £5,000 and over, 41 per cent of the wives (of all ages) had professional earnings and profits. In the income range £2,500–£4,999 the proportion was 44 per cent for 128,000 units. For all married couple units above the exemption limit (11,923,000 including those in the ranges £2,500 and over)the proportion was only10 per cent.[3]

(*e*) More units reduced in value as a result of the operation of irrevocable settlements, discretionary trusts, family and educational trusts and gifts *inter vivos* in favour of children, grandchildren and other kin or 'life tenants' by parents, grandparents and other relatives by blood and marriage.[4] These processes are also likely to create among child and other beneficiaries below the surtax level more single income units. The splitting and spreading of top incomes over time and on a kinship basis probably causes more bunching of incomes in the distribution as a whole. The object of giving as much income as possible to each member without attracting higher rates of tax is likely to have this effect. Consequently, some units will be raised from lower ranges; others will be newly created; for example, more children and young people, previously nil income-receivers or whose small incomes were aggregated for statistical purposes with parental incomes, appear as middle-range income units in their own right. Some evidence in support of this thesis is to be found in the study of 'The Distribution of Personal Wealth in Britain' by Messrs Lydall and Tipping.[5] One of their significant conclusions, in analysing estate duty returns for 1951–6, was that the greatest degree of con-

---

[1] See statistics on p. 60.

[2] There is no evidence that in their yearly statistics the Board of Inland Revenue 'marry' these separate assessments (often involving different schedules and district office files). This would, in any case, be a difficult routine administrative operation because many of these high income wives have professional names and are often 'incorporated'.

[3] See p. 59.

[4] Developments in these fields since 1938 are discussed in Chapter 5.

[5] Lydall H. F. and Tipping D. G., *Oxford Inst. of Statistics Bulletin*, 1961, Vol. 23, No. 1, p. 96.

centration of wealth is to be found amongst the youngest adults (age 20–24). While the average net capital held by those with more than £2,000 varied from nearly £16,000 per person for those aged 20–24 to less than £10,000 per person for those in the age groups 45–74 (and about £11,000 for the 75's and over), the average for all holdings ranged from £330 per person in the group 20–24 to £2,310 for the 75's and over. The evidence in later chapters suggests that this deliberate redistribution within families of large fortunes has markedly increased since 1938.

(*f*) More units reduced in value as a result of deliberate reductions of high salaries to spread income into retirement and pay less tax over the life span as a whole.

(*g*) More units reduced in value so as to acquire the right to tax-free lump sums of up to £40,000 or more on actual or nominal retirement.

(*h*) More units reduced in value as a result of higher expense accounts compensating for lower salaries, fees and earnings.

(*i*) More units reduced in value as a result of some part of fees, salaries and earnings being translated into or exchanged for income in kind (not taxable or taxable at lower values) for the earner, his wife and/or children and other relatives.

(*j*) More units reduced in value as a result of lower fees, salaries and earnings being compensated with bonus shares carrying tax-free capital gains and tax-free gains on retirement.

(*k*) More units reduced in value as a result of compensation for real or contrived severance of employment thus transforming income into capital.

Again, this list by no means exhausts the subject of the ways in which current income may be split or spread or transformed into capital for oneself or for others. In later chapters we bring together some of the evidence in support of these propositions relating to 'the arrangement of personal income'. Much of this evidence applies with particular force to the top income ranges.

We cannot, in this review of tax population estimates and of the social and demographic changes which have affected the total population and the composition of different parts of it, reach any statistical conclusions. It is impossible to do so in the absence of the necessary data for the years since 1938.

Nevertheless, as a general statement it can be said that many profound changes have been at work some of which have given the Board's tables a spurious egalitarian look. Others have been genuinely equalizing in their effects. Others again have worked in the opposite direction and have affected the top and bottom income ranges differently. The genuine effects as distinct from the spurious ones cannot be disentangled.

These problems arise in part from the use of annual or 'snapshot' pictures of income distribution. While such statistical pictures have their uses for some purposes they cannot be relied on for others. Not only are they faulty in many respects for all purposes but, in addition, they have been expected or asked to carry broader conclusions which, over the years, they have been increasingly incapable ef supporting.

One further point remains to be made before we turn to the vexed question of the income unit. The discussion in this and preceding chapters has all been conducted in terms of income *before* tax.[1] Had we taken instead incomes *after* tax most—if not all—of the criticisms entered against the interpreters of income distribution statistics would have to be more heavily emphasized. As income is spread, split, reduced or transformed into capital more personal and other allowances may be attracted or the proportionate weight attached to these allowances in relation to gross income may be enlarged. Many of these procedures are, indeed, so designed as not only to reduce the immediate incidence of tax but to attract more allowances and to enhance the value of existing ones.

Although we criticize, with some severity, those who have handled the official statistics incautiously, we need to remember and re-echo at this point the words of W. J. Ashley: '. . . one reason why the very considerable speculative ability of English economists has so often the appearance of beating the air, *bombinans in vacuo*, is that it has not had furnished to it sufficiently concrete statements of actual groups of circumstances.'[2]

---

[1] Mr Lydall claims that 'the direct effects of taxation bear comparatively little responsibility for the reduction in inequality of post-tax income' over the period 1938–57 (*op. cit.*, p. 8). Professor Paish makes much the same point (*op. cit.*, p. 14). On the other hand, the Bow Group in *Taxes for Today* (1958, p. 27) concludes that 'The change since 1938 in the distribution of incomes *after tax* (their italics) demonstrates the virtual extinction of the pre-war "surtax class".'

[2] Ashley W. J., *The Adjustment of Wages*, 1903, p. 7.

Note on 'percentiles'.

The percentiles divide the income units, arranged in order of size, into 100 groups each containing an equal number of units. Thus in Table 1 (p. 38) for 1938 the first percentile value is £1,140, which means that 1 per cent of all incomes were greater than this and 99 per cent smaller.

CHAPTER 4

The Individual and the Family

I

IN what may seem to have been an illogical order we have discussed first the question of the income population and its division into parts before considering the nature of the units that make up the whole. But this arrangement was thought to be helpful in explaining why a detailed analysis of the unit problem is essential. Having looked at the universe—and found it nebulous—we may now inquire into the inhabitants.

It is obvious enough from the preceding chapter that the official statistics of income are presented neither in terms of the individual nor of the family. From the steps we have so far taken in this analysis it appears that they embody a mixture of both. But in what respects and to what extent has the mixture changed since 1938? And what is the pattern for different income groups and in what direction has it been changing?

These questions are crucial for understanding what we mean when we say that the distribution of incomes has become more (or less) unequal. Are we talking about individual Robinson Crusoes, households, heads of households, nuclear families, husbands and wives or wider kin groups? These questions, surprisingly enough, are seldom if ever asked by those who study the Board of Inland Revenue's data. They accept the Board's presentation which, understandably enough, reflects the basis of the system of direct taxation in the United Kingdom. That is not to say, however, in a better organized statistical world that they need necessarily do so. The obstinate fact remains, though, that they have always done so. Perhaps this mattered less in the nineteenth century when income tax was not a mass tax and economic theory was peopled with Robinson Crusoes without progeny. But we know more today about the realities of economic behaviour, the structure of family relationships, and the significance of age and other characteristics in respect of the property rights of earners and non-earners alike.

This is not the place to dissect all the ramifications of fiscal law concerning the unit of taxation. 'Family,' said the Royal Commission on Taxation in its *Final Report*, 'is not a word of precise significance,'[1] having come to the conclusion, in an earlier report, that the incomes of parents and children should not be aggregated, apparently because there was some doubt as to whether children shared the same standard of living as their parents.[2] An additional reason was that no evidence was put before the Commission indicating that considerable numbers of children were being placed in possession of substantial incomes during the lives of their parents. The Commission, passively contemplating a somewhat mythical family, do not seem to have actively sought such evidence—a point we shall return to in later chapters.

No recommendation was therefore made for any change in the law which, as a rule, does not require the child's income to be aggregated with that of the parent. The Commission were in favour of maintaining the general rule of aggregation for the incomes of husband and wife. They pointed out, however, that this rule had been departed from to a considerable extent by various enactments which gave favourable treatment to the earnings of wives. 'It might well have been expected that the introduction of this régime' (special relief for wife's earned income, the benefit of the reduced rate reliefs, and the breaking of the aggregation rule up to a certain limit) 'would have been accompanied by the withdrawal of the marriage allowance from the husband, for there seems to be little logic in conceding to married persons both a marriage allowance for one and a separate single allowance for each. By this device they receive treatment not less but more favourable than if each was a single person.'[3] Accordingly, the Commission recommended some reduction of the special reliefs given to the married couple where both are earners.

No action has been taken by the Government since the Commission reported in 1954 to implement this recommendation. We need not recount here the reasons; it is sufficient for our purposes to draw attention to the fact that there are taxation advantages in the case of two married earners, and that these advantages may result in more requests being made for the separate assessment of both incomes. The reasons would be simply administrative ones, as no tax saving would follow. Nevertheless, these married earners might then appear in consequence as separate units in the annual income statistics. More separate returns and assessments may thus complicate the statistical picture because of the difficulties, referred to earlier, of 'marrying' information from different sources for the purposes of statistical presentation.

[1] Cmd. 9474, 1955, p. 54.
[2] *Second Report*, Cmd. 9105, 1954, p. 38.
[3] Cmd. 9105, p. 39. See also Plunkett H. G. S. on 'The Taxation of Married Women', *British Tax Review*, June and September, 1957.

II

In its latest report for 1960 (and in earlier reports as well) the Board, in an explanatory note to its statistical tables, writes in terms of 'individuals' and 'individual persons'.[1] This note is followed by a table (table 28) giving the 'Estimated number of individuals with total incomes above the exemption limit'. Many readers have naturally been led to infer that the statistics relate to biologically separate individuals.

In its report for 1958, which provides additional tables on the 1954–5 income survey, the introductory text refers to 'personal incomes' and the 'number of incomes'.[2] Later tables in this report (e.g. table 68) are headed 'All Persons'. The totals for this table; for tables headed 'number of individuals', and for tables headed 'number of incomes' all correspond precisely. The totals for other tables (tables 64–7) which give separate data for 'self-employed persons, employed persons, pensioners and investment incomes', when added together, furnish exactly the same figure. In the first table ever published by the Board (for 1918–9) on the distribution of incomes the income unit is stated to be the 'individual'.[3] The signposts do not seem to have become more helpful over the years.

However, we learn elsewhere from some of the Board's reports which, from time to time, provide an outline of the taxes in force, that in assessing tax 'the incomes of husband and wife are aggregated and are treated as one income'.[4] That this principle is applied to some of the statistics is made clear in the reports which classify Schedule E income. We might assume, therefore, that the Board, in presenting its tables, uses the word 'incomes' in this sense, and 'individuals' and 'persons' in another. But this is not so. For example, there is one table in the 1958 report (table 59) from which we learn that in 1956–7 there were 8,900,000 single persons (including widowed) and 12,000,000 married couples, a total of 32,900,000 individuals. Included in the 'single' category, there were 1,330,000 single persons with children and/or dependent relatives and/or resident house-keepers. Of the 12,000,000 married couples, there were 6,606,000 with children and/or dependent relatives and/or resident house-keepers. As we shall see later, an unknown proportion of the children and, maybe, other 'dependents' of the married couples and single persons appeared, in addition, as single persons. The 'Total Single and Married' in this table is given as 20,900,000. This is precisely the same as that given elsewhere for 'individuals', 'all persons', and

---

[1] *BIR 103*, pp. 28–9.
[2] *BIR 101*, pp. 69–72.
[3] *BIR 63*, p. 70.
[4] *BIR 101*, p. 3.

'number of incomes'. The terms are clearly synonymous. The definitions employed can thus be deduced, but one has to be skilled and
industrious to find out.

As no information to the contrary is given in any of the Board's
reports, it would therefore seem logical to assume (*a*) that *all*
married couples whether living together or separated for any number
of reasons are treated statistically—if not always for tax purposes—
as one unit[1] (*b*) that *all* married couples where the wives have a
separate source of income which in value is *below* the exemption
limits are similarly treated (*c*) that *all* divorced persons are treated as
single persons and (*d*) that *all* married couples where the wives have
separate incomes (earned and unearned, above and below the exemption limits) and ask to be separately assessed are, nevertheless,
treated statistically as one unit.

But are these assumptions correct? How does the Board deal
statistically with marriages that are broken within a tax year?[2] How
does it in practice classify and present the data for husbands and
wives in cases where the wife (or husband) elects to be separately
assessed for professional, business or social reasons? There may be
1,000 or 100,000 such cases. In theory, the great majority could be
concentrated among the top 5 per cent of income groups. Nor is it
possible to tell how the Board handles statistically the incomes of
adult individuals who are partly dependent on other tax units.

These are some of the questions which confront the assiduous
readers of the Board's tables. In answer to the questionnaire we
submitted, the Board said that our assumptions were correct (see
Appendix A); all married couples, whatever their circumstances,
however large or small the income of the wife, and irrespective of
residence, different sources of income, elections to be separately
assessed, and the use in assessment of different 'professional' names,
are counted, in all census and non-census years, as one unit. But this
we should remember is what is supposed to happen. In actual fact,
as we shall see later, the counting is far from complete; for instance,

---

[1] The parties to a marriage who are separated are, however, treated as single
persons for tax purposes (see Income Tax Act, 1952, S. 361 (1) and also the
discussion on 'Husband and Wife' in the *Report of the Income Tax Codification
Committee*, Cmd. 5131, 1936, pp. 80–4).

[2] Although the Board gives no information, the Central Statistical Office in its
adjusted income distribution tables counts a woman who is single or divorced for
part of the year as having a separate income of the amount received while she
was single or divorced. 'For this reason, and also because of deaths which occur
during the year, the number of small incomes is larger than it would be if incomes
were measured by their annual rates at a given point in time.' (*NIBB*, 1959,
p. 69.) Differential mortality, divorce and separation rates by income class,
family size and over time must clearly introduce, therefore, a further problem of
bias in the income distribution statistics.

there were something like 1,500,000 'missing' earning wives in the sample survey data for 1954–5.

Moreover, as the Board also confirmed, we do not know how many women—and how they are distributed by income ranges—appear twice in the same year in the Board's tables; once as single, widowed or divorced units and once as married couple units. Thus we do not know how much double-counting has occurred, nor where it has occurred, in each year since 1937–8. This may be an example of a spurious unequalizing effect; 'the number of small incomes is larger than it would be if incomes were measured by their annual rates at a given point in time'.[1]

Such factors as these, and their precise statistical treatment over the past twenty years, could be of substantial importance in spuriously altering the shape and distribution of incomes, particularly in the top ranges. This possibility is not mentioned by Professor Paish, Mr Seers, Professor Cartter, Professor Allen, Dr Rhodes, Messrs Shirras and Rostas, the Bow Group Report and other studies of the size distribution of incomes. Mr Lydall does draw attention to the fact that the marriage habits of the population and 'changes in the proportion of married women who go out to work' will affect the shape of the income distribution.[2] He does not, however, discuss the other factors to which reference is made above. Moreover, he says that it is the Board's practice to include all dependents in the tax units of either single persons or married couples.[3] Quite apart from the extremely difficult problem of defining a dependent for these particular statistical purposes this, it would seem, is unlikely to happen in practice in all cases.

Mr Worswick, the most sceptical interpreter of personal income statistics, pointed out that the 'individual' is not 'sharply defined since for income tax purposes (and it is from the Inland Revenue, that many of the data are obtained) the incomes of husband and wife usually, though not always, constitute one undivided income, while other data used, especially for the lower incomes, attribute one income to one person, married or not'.[4]

Professor Cartter, in mentioning that husbands and wives 'if filing a joint return, are counted as a single income'[5] allows it, quite

---

[1] *NIBB*, 1959, p. 69.

[2] Lydall H. F., *op. cit.* (1959), pp. 6 and 22–3.

[3] Mr Dudley Seers came to the conclusion, however, that 'the incomes of other dependants are not included' (Seers, D., *op. cit.*, p. 32). This is a further example of the many confusions about the practices of the Board in its treatment of income data which emerge when a comparison is made of the definitions employed by different income statisticians.

[4] Worswick G. D. N., 'Personal Income Policy', in *The British Economy 1945–50* (ed. G. D. N. Worswick and P. H. Ady), 1952, p. 314.

[5] Cartter A. M., *op. cit.*, p. 28.

reasonably, to be assumed that if separate returns are filed the persons concerned will be treated as two separate units in the statistics. This would be a sensible conclusion in view of the known difficulties of 'marrying' information from different local offices about different individuals, sometimes using different names and often taxable under different schedules.

Among the reasons that could lead to separate statistical treatment in some of the Board's tables (because of the double-counting problem, separate returns and so forth) two may be briefly mentioned. For wives assessed under Schedule D in respect of 'profits and professional earnings' and entitled to expense and other allowances there may well be, in many cases, distinct advantages for accounting purposes in separate returns and assessment. The proportion of wives in this category rises very sharply with income as the following analysis shows. This is taken from the Board's classification of the earnings of 2,911,500 wives in 1954–5[1] (the latest published figures):

TABLE 4

WIVES' EARNINGS 1954–5

Range of profit or remuneration (wives only) £	Total number of wives	% with profits and professional earnings	% with wages and salaries
165– 499	1,907,000	4	96
500– 799	96,600	16	84
800–1,499	13,100	50	50
1,500 and over	3,800	61	39

The difference in the two distributions is clear enough, though how this affects the distribution by income ranges for all income units it is impossible to estimate. It could imply marked differences by income range in statistical treatment and in the handling of various difficulties of classification and presentation. Moreover, it should be borne in mind here that the Board discovered in classifying its sample survey data for 1954–5 something like 1,500,000 'missing' earning wives (a problem of bias which is discussed below). We do not know what proportion of these missing wives (and how they were distributed by married couple income unit ranges) were missing because of separate returns and assessments and other administrative practices.

A second, and probably more powerful, factor in considering potential distortions in the income distribution tables is to be found in the different proportions of wives with earned incomes. An analysis of the Board's figures classifying incomes by type and derived from the 1954–5 sample survey shows:

[1] *BIR 101*, table 63.

TABLE 5

MARRIED COUPLES 1954–5:

PROPORTION WITH EARNING WIVES[1]

Married couple income range (net income before tax) £	Numbers	% with wife's earnings (earned income only)
155– 499	4,837,000	7
500– 699	3,963,000	22
700– 999	2,156,000	45
1,000–2,499	796,000	38
2,500–4,999	128,000	36
5,000–9,999	34,000	25
10,000 and over	9,000	17

In addition to the marked and surprising differences by income classes in the number of income units with wives earning, what is particularly interesting about these figures is the substantial proportion of wealthy taxpayers with earning wives.[2] The point will be further discussed later. Meanwhile, it may be said that the classification of married couples as single tax units can give a misleading picture of the distribution of individual incomes when there are such large differences in the proportions of tax units which comprise two earners.

It has, however, to be remembered that in the 1954–5 sample survey there were these 1,500,000 'missing' earning wives—missing from the above classification. Only 2,562,000 earning wives appeared in the Board's returns for taxpayers above the £155 exemption limits.[3] Other estimates derived from the Ministry of Labour and various sources of all wives earning for any part of the year resulted in a figure of over 4,000,000 for 1954–5. Some of the missing 1,500,000—though for obvious reasons only a relatively small number—would have fallen in the married couple income unit class below £155. Their absence may in general though have depressed the incomes of the bottom groups. The error appears to have been due to the omission or understatement of wives' earnings from or on reports made by the husband's tax district. According to the Board, these reports 'were based on the husband's personal return (supplemented in many cases by information passed by the wife's district)

[1] Schedules D and E combined. They are not separated in the relevant table (table 58, *BIR 99*). These are the latest published figures.

[2] Under Section 142 of the Income Tax Act, 1952, where one business is carried on by the husband and another by the wife, a loss on the business carried on by the one may be set off against the profit on the business carried on by the other. It is impossible to say how this provision affects the income distribution statistics. It could be important in the married couple income range above £2,500.

[3] A later and still incomplete classification of wives' earnings in 1954–5 arrived at a total of 2,911,500 (*BIR 101*, table 63).

and should therefore have included the wife's wage or salary even if below the tax deduction card limit'.[1]

These errors in and omissions from the income distribution statistics appear to have grown over the years. In 1937–8 214,000 earning wives (Schedules D and E[2]) were reported in the census; nothing was then said about missing wives. In the 1949–50 census the number reported rose to 1,412,000 (Schedule E only[3]) when it was said that there were 'over a million married women in paid employment' missing.[4] In the 1954–5 sample survey, as we have seen, the number was 2,562,000 (Schedules D and E) and the number of missing wives around 1,500,000.

In referring to this last figure, the Board stated that 'No loss of tax was involved' because of deductions under PAYE (Schedule E). The Board's 'approximate estimate' of the wives' total earnings (including earnings below the tax deduction card limit) was 'over £800 million'.[5] As the total which actually appeared in the Board's report on the 1954–5 sample survey was £560 million[6] the deficiency in respect of the classification of 'wife's earnings' was over £240 million. The statement that 'no loss of tax was involved' must imply that PAYE was paid on the wives' earnings at a rate which took account of the husbands' income. It follows, therefore, that the missing 1,500,000 wives were assessed under Schedule E and presumably appear as single women in the statistics. If they appeared as married, they could not be missing.

On some unexplained basis, which may or may not have removed the inflation of single tax units, the Board made adjustments to these 1954–5 data to show 'the estimated distribution of incomes by ranges of income before and after payment of tax which would have emerged if the reporting had been complete'.[7] Nor is it clear what adjustments were made in this year and other years for the fact that the Board employs different definitions of income for men and earning wives in its classified tables. For example, income before tax for men is after the deduction of the National Insurance tax; for earning wives this tax is not deducted; nor are superannuation contributions and 'other allowable expenses'.[8]

[1] *BIR 101*, p. 70.
[2] This is what table 21 (a) *appears* to imply, but it is only an assumption as no detailed information is given (*BIR 83*, p. 32).
[3] No separate figures were given for Schedule D in 1949–50. In 1954–5 the number 'reported' under this Schedule was 97,100 plus an 'incomplete' figure of 40,500 for those below the exemption limit (*BIR 101*, pp. 70–1 and 78).
[4] *BIR 94*, p. 96.
[5] *BIR 101*, p. 70–1.
[6] *BIR 99*, table 55.
[7] *BIR 99*, p. 87.
[8] See also p. 103.

The Board's adjusted tables then became the key tables for all the analysts who have made comparative studies of income distribution. Two years later, the Board, in publishing additional tables on the 1954–5 data concerned with the classification of incomes by range of earned and investment income, said that it was unable to carry these adjustments into the additional classifications.[1] We cannot tell why this was impossible if we do not know what distributions were assumed for the missing 1,500,000 wives.

This is not the only 'distribution pattern' which is applied by the Board on some unexplained basis to its statistics in census and non-census years. In non-census years, the Board appears to apply a distribution of incomes pattern, derived from the latest census, to its current statistics of individual schedule assessments. It did so, for example, in 1948–9 when the 1937–8 distribution was used as the basis for the classification of 1948–9 assessments.[2]

At the time of writing (summer 1961) the latest published figures of the size distribution of incomes relate to 1958–9. These are based on the statistics of income charged; on 'projected' 1954–5 income distributions, and 'projected' 1954–5 'adjustments' in respect to the missing 1,500,000 earning wives and other deficiencies in the 1954–5 sample survey which are discussed below. The combined effect of these different assumed 'distribution patterns' on the income statistics by range and type could be quite profound. But how important it might be cannot be estimated in the absence of any information about the bases of the assumptions.

The last chapter drew attention to the far-reaching changes which had taken place in the social and demographic structure of the population since 1938. It becomes even harder, faced with the problems of statistical analysis and presentation so far described in this chapter, to assess the impact of these changes on trends in the size distribution of incomes in terms of either the individual or the family. Any attempt to reconcile the Board's population figures with those of the Registrars-General has to be ruled out completely.

III

It is clear that the Board's definition of an income unit is not a 'family' definition. But how has it treated since 1938, in its income statistics, the incomes of children and adult dependents? We have been unable to discover in any detail in any of the Board's Reports for at least the last fifty years how the income of dependents, and

---

[1] *BIR 101*, p. 69.
[2] *BIR 92*, p. 82. In 1950–1, which 'projected' the distribution of 1949–50, 'adjustments' were made for the one million 'missing' earning wives in 1949–50 (*BIR 95*, p. 92).

especially the income of child dependents, is classified and presented in the Board's tables. Nowhere is a child or an adult defined. At what age, if any, does an individual become an income unit? When it has any income or wealth (however defined) or only when it 'earns' more than a specified amount?

According to the Common Law in England a person is an infant, a child or a minor until he comes to full age at twenty-one; and since, in Halsbury's words, an infant is regarded as 'of immature intelligence and discretion' his capacity to bind himself by contract is closely limited. The Income Tax Act, 1952, embodies this definition, and includes special provisions to safeguard the interests of infants in regard to settlements and trusts.

Nevertheless, age twenty-one does not seem to be significant in the classification of the income distribution statistics. This problem is, of course, a part of the much wider question raised by the complete absence of any age classifications in the official statistics. Among all the social and demographic changes since 1938 surveyed in the preceding chapter this is probably the most important single gap in the Board's statistics, and one which critically affects the conclusions that have been drawn from them. In comparisons over time, it is as necessary to take account of the age of the unit as it is to be clear about the definition of a unit and the concept of income.

The fact that, in fiscal law, the incomes of children are not aggregated with the parents' incomes makes the questions of age and statistical treatment even more important in a period when the status of children, social and economic, has been rapidly changing. Some evidence concerning these changes is given in later chapters. So far as the higher income groups are concerned, they are reflected in the finding by Messrs Lydall and Tipping that the greatest degree of concentration of wealth by age is now located among young adults.[1]

As no information about the statistical treatment of children since 1938 can be elicited from the Board's tables, we must consider what other official and unofficial reports have made of the question. The latest National Income Blue Book states that the 'income of a juvenile, even if he is partly dependent on his parents, is counted separately'.[2] As the relevant tables do not include incomes under £50 in the year, it may be assumed that all children under the age of 21, with incomes (earned and unearned) over £50 in a year, are counted in the Board's statistics and in the adjusted Blue Book tables as separate tax units.[3] The Board, while confirming that the income of children is not

[1] See pp. 51–2.
[2] *NIBB*, 1960, p. 69. A similar explanation is given in *National Income Statistics: Sources and Methods*, 1956, p. 70.
[3] The Blue Book does not refer to the incomes of other dependants.

aggregated with that of their parents,[1] is nevertheless unable to provide any estimate of the number of minors' incomes—earned and unearned.

Different interpretations have, however, been drawn by those who have used the Board's statistics. Dr Barna's analysis of the 1937 material was based on the assumption that the incomes of children under 16 (or over 16 if continuing education) were counted as part of the income of their parents. 'Earning' children under 16 were included; those above 16 were excluded. He also assumed that the incomes of dependent adults incapacitated by old age or infirmity were similarly treated.[2] The study of Messrs Shirras and Rostas for 1937–8 and 1941–2 followed this interpretation with one difference. They stated that children outside these categories were counted with 'parent units' when the children had only 'insignificant, independent incomes'.[3] The studies by Dr Rhodes of the 1937–8 and 1946–7 material,[4] by Professor Allen of higher incomes in 1949–50 and 1954–5[5] and by Professor Paish of the statistics for 1937–8, 1949–50 and 1953–5,[6] make no mention at all of this problem of the income of children and other dependents. In Mr Dudley Seers' study *The Levelling of Incomes since 1938* it is stated that, in British income tax practice, 'the incomes of other dependents are not included' in the incomes of husbands and wives.[7]

In his book *The Redistribution of Income in Postwar Britain*, Professor Cartter was obviously troubled by these questions. However, he concluded that children (aged over and under 16) and adults 'at least partially dependent upon others, with incomes of less than £120, are classified as dependents in these tax families'. No distinction was made between earned and unearned incomes. Professor Cartter assumed that this definition was operative in 1948–9.[8]

In his most recent study of the size distribution of incomes, Mr Lydall makes no reference to the income of children beyond a com-

[1] Apart from one exception. The income of a child is aggregated with his parents only when it arises from a deed of covenant or other provision made by them that has been disregarded for tax purposes. Scholarships and education grants for full-time instruction are not taxable and are not therefore counted as income.

[2] Barna T., *Redistribution of Income through Public Finance in 1937*, 1945, pp. 63 and 254.

[3] Shirras G. F. and Rostas L., *The Burden of British Taxation*, 1942, p. 3.

[4] Rhodes E. C., 'Distribution of Earned and Investment Incomes in the United Kingdom in 1937–38', *Economica*, New Series, 1951, No. 69.

[5] Allen R. G. D., 'Changes in the Distribution of Higher Incomes,' *Economica*, New Series, 1957, No. 24.

[6] Paish F. W., *op. cit.*

[7] Seers D., *op. cit.*, p. 32.

[8] Cartter A. M., *op. cit.*, p. 24.

ment that 'there has been some redistribution of capital from the old to the young, in order to avoid death duties, which results in a less unequal distribution of investment income'.[1] In his earlier book, Mr Lydall stated that, from the point of view of the Inland Revenue, children with subsidiary earnings are separate income units: 'their income is not included in the income of their parents and, if they earn more than £135, they are separately assessed'.[2] No mention was made of unearned income.

These examples and quotations show how much obscurity has surrounded the whole matter. What is at stake is the statistical treatment in the income distribution tables since 1938 of the entire population of single dependent or partially dependent children aged under 21 (and presumably older students if they are continuing their education full-time and continuing their 'dependency') and the population of adults, incapacitated by old age or infirmity, who are wholly or partly dependent on other tax units. Quite apart from all the social and demographic factors of change discussed earlier, any differences in the administrative and statistical handling of these two populations since 1937–8 could have a marked effect on the shape and pattern of the distributions in question. There may have been changes in statistical treatment; this might explain in part the variety of statements about the matter.

Very little is known or can be gleaned from other sources about trends in the standard of living of these two populations. There is some evidence to suggest, however, that the standard for children and young people in general has risen markedly since the 1930's with the virtual abolition of juvenile unemployment and, indeed, it may well have risen in relative terms to a greater extent than among the rest of the population. The mixing up of minors' incomes as separate statistical units (if that has been the case since 1938) with married couple units could have had equalizing effects on the total distribution if more children have had unearned incomes and if their earnings have risen more significantly than those of adults. Or the reverse effect may have operated, depending on the changes in the number and sizes of such incomes. What are spurious or genuine equalizing effects (or the reverse) cannot be defined and disentangled without far more information than we have at present.

If the facts were available, the differential effects of social class, sex and source of income among minors on the whole question of inequalities of income could well repay study. It may be that a greater concentration of wealth among young people today will result in a more unequal distribution of incomes among adults in the future. This possibility is raised by the discussion in later chapters

---

[1] Lydall H. F., *op. cit.* (1959), p. 22.

[2] Lydall H. F., *British Incomes and Savings*, 1955, p. 18.

E

concerning the manipulation and rearrangement of incomes and wealth in favour of children by means of a variety of settlements and trusts.

What is clear from all the evidence is that the Board of Inland Revenue has no knowledge whatever about the current incomes of children under 'accumulation trusts'. Income accumulated for a child under the age of 21, to which he becomes contingently entitled on attaining that age, is not his income in the year in which it is received and so he cannot be assessed to surtax on it.[1] When he does attain 21 he gets the accumulations as capital so they never get assessed for surtax. Nor has the Board any data on the number and sizes of minors' incomes.

There may be other sources of income, earned and unearned, of minors which are not separately classified by the Board because they are not distinguished in parental tax returns or for other reasons. It is difficult to see how all these items may be handled statistically, irrespective of whether they are or are not subject to tax. Just as the Board has discovered in recent years a large number of missing earning wives so it has also discovered many missing children.

In the Board's 92nd Report, dealing with the classification of incomes in 1948–9, mention was made of the fact that 'the number of dependents may be underestimated because of failure by persons not liable to tax to claim all the allowances due to them'.[2] After the special income census had been taken in 1949–50 the Board reported about 500,000 missing children. The same explanation was offered, and the Board added that it had no information 'on which to estimate the correct distribution of the missing children'.[3] In the following year, however, and somewhat surprisingly, adjustments were made to the 1949–50 data on some unexplained basis to cover— what the Board described as—'the omission of *some* (our italics) children from the 1949–50 Income Census Reports'.[4] When the results of the 1954–5 sample survey were published the Board reported: 'The third discrepancy is that the number of children for whom allowance is claimed is about 250,000 less than would be expected from the population estimates made by the Registrars-General and the figures of children continuing their education beyond the normal school-leaving age. There is a corresponding deficiency in the number of cases with family allowances. These omissions, which may arise from a number of causes, do not imply any corresponding errors in the amount of tax collected.'[5]

[1] *Stanley v. IRC* (1944), 1 K.B. 255.
[2] *BIR 92*, p. 86.
[3] *BIR 94*, pp. 96 and 117.
[4] *BIR 95*, p. 91.
[5] *BIR 99*, p. 87.

It seems probable that these deficiencies are, as the Board concludes, 'due mainly' to failure by taxpayers to claim a further allowance when another child is born in cases where there is in any case no liability to tax.[1] But it is unlikely to be the explanation in a minority of cases. Where the child has income in his own right exceeding £100 for the year (excluding scholarship income) a child allowance cannot be claimed by the parents. The Board has no information about the number of such cases and their distribution by parental tax unit incomes since 1937–8 or before.

The treatment of children, married women, separated spouses and adult dependents are not the only questions raised by this problem of defining income units. Another is the transformation of individuals, persons, cases or income units into companies by incorporation. Families and kin groups can similarly be transformed. In a later chapter we discuss the substantial rise in the number of 'one-man companies' in recent years. This may represent yet another factor in changing or distorting the picture we have of the distribution of incomes by ranges, particularly at all levels over £2,000 a year.

We may fittingly bring this frustrating chapter to an end with the observation by Messrs Blum and Kalven that, in any system of progressive taxation, 'there seem to be no criteria for deciding whether the unit ought to be the individual, the marital community, the family, the household or some other combination of persons'.[2] That being so, we should recognize that inconclusively defined statistics, which relate wholly neither to individuals nor to families, have little value in themselves.

[1] *BIR 99*, p. 87.
[2] Blum W. J. and Kalven H. Jr., *The Uneasy Case for Progressive Taxation*, 1953, p. 16.

CHAPTER 5

Tax, Time and Kinship in the Arrangement of Income

I

WE have discussed the universe and the individual. The first we found to be nebulous; the second insubstantial. To complicate matters further, we now have to consider the relativities of time and kinship.

First we must draw attention to some of the assumptions implicit in the conventional analysis of the size distribution of incomes. An individual's consumption—that is, his standard of living—is assumed to be related to his statutory earnings or income in an arbitrary period of a year. For purposes of statistical measurement, this assumption has necessarily been applied without distinction to different income classes who receive their income on a different time basis—weekly, monthly, yearly or in respect of longer periods. The individuals who constitute these classes at different points in time have, in consequence and for various historical and cultural reasons, different sets of attitudes, behaviour and propensities in relation to getting, spending and hoarding. They are conditioned, as Keynes said, by different anticipations of the future and by 'all sorts of vague doubts and fluctuating states of confidence and courage'.[1] They include those, like old age pensioners, whose command over resources-in-time is often limited to a week; professional people whose expectations and behaviour are influenced by the concept of 'career earnings';[2] the business man and the self-employed with fluctuating degrees of confidence and disquietude about money and

[1] See his chapter in *The Lessons of Monetary Experience* (Ed. A. D. Gayer), 1937, p. 151.

[2] Notably the doctors. See the discussion of career earnings in the *Report of the Royal Commission on Doctors' and Dentists' Remuneration* (Cmd. 939, 1960, pp. 39–40). The Commission appear to have disregarded this concept in their conclusion that social changes since 1948 'greatly narrowed the gaps between the standard of living of the whole of the professional classes and that of the wage-earning majority' (p. 107). The Board of Inland Revenue's income statistics obviously played an important role in the deliberations of the Commission.

time; and many other functionally different groups with varying propensities to consume and to save.

There is no need to labour the point. What we wish to underline are the possible effects on the income distribution statistics caused by different degrees of command over resources-in-time by different income classes. Any definition of 'resources' would, to be realistic, have to include much more than income actually received and statutorily recorded for a specified period. It would have to embrace the possession of stored wealth, command over certain other people's income-wealth, expectations of inheritance from the past and un-taxed gains in the future, power to manipulate and use the critical educational, occupational and nuptial keys to wealth advancement, and much else besides.

Here we are only concerned with a limited part of the wider notion of a *masse de manoeuvre* in command over resources-in-time. This we may conveniently describe as the power and opportunity to rearrange income-wealth over time. Nevertheless, while employing this more restricted notion, we should recognize that no discussion of equality in the twentieth century can be adequate unless it takes account of a wider definition of command over resources-in-time.

The degree to which individuals have the knowledge, the opportunity and the expertise to rearrange and spread their income over time varies greatly. These characteristics are much more likely to be found among those at the top of the conventional income distribution table than among those at the bottom. In Chapter 7 we shall show that this is true in respect of the spreading of income into the years of 'retirement'. Rational anticipations of a longer and more certain span of life after work, accompanied no doubt in many instances by a deepening disquietude about the texture of the life to be lived in these added years, have clearly played a part in the growth of arrangements for retirement spreading.

The 'smoothing' of income over the working years and its spread into retirement is not, however, the only motive for rearranging one's income. Others are discussed in this chapter. The one characteristic they all have in common—hence the title of this chapter—is that they have their source in social relationships. Income-wealth is rearranged over time on a family or wider kinship basis. A different conspectus of time and a different view of economic man as a taxable unit are thus introduced to complicate the statistical problems of measuring inequality.

If we could assume that what was true in 1938 in respect of all these factors which enter into the measurement of the size distribution of incomes was similarly true of the 1950's then our problems would be more manageable. But we cannot. The emphasis we have given throughout this study to social changes, and the evidence brought together in earlier chapters concerning the role of demographic and social forces of change, dispute the assumption. It is a tenable thesis that these forces which, as we have shown, influence in many complex ways the range and pattern of income distribution—particularly at the higher levels—have also been accompanied by developments in the means by which income-wealth may be rearranged. The factors which helped to bring about these changes in family life and social structure since the 1930's may have themselves contributed to shaping different attitudes towards time and kinship in relation to the disposition of income and wealth. Such changes in attitudes may, in their turn, have led to a wider variety of methods by which income-wealth is rearranged.

II

In this chapter we are chiefly concerned with those acts or methods which, for want of a better term, we have described as 'manipulative' to distinguish them from other forces which affect the pattern of income distribution. They include, but extend much beyond, those which are commonly thought of as tax avoidance and tax evasion. In greater or lesser degree, though, they involve an act of deliberate forethought in the 'arrangement' or 'rearrangement' of income and wealth. They take many forms, some so complex as to defy the wit and ingenuity of a Torquemada. No attempt will be made, however, in these pages to provide a comprehensive survey. The material is drawn from many sources and, as will be seen, is of varying value. In a few instances, certain data are presented which help to quantify the factors studied. In general, however, the most that can be done is to show in descriptive fashion something of the order of magnitude and the scale of growth of these 'manipulative' factors considered as a whole.

As a number of specialists in fiscal law have pointed out, the United Kingdom is remarkable in the Western world for the generous legal opportunities it allows for the alienation of income by means of irrevocable covenants. There are many types of covenants and settlements *inter vivos* which allow the income and wealth of an individual to be spread and shared among a large number of other individuals related by blood or marriage. There are also many ways by which taxable income may be transferred and 'given away' to individuals who are not related to the covenantor.

Most of these arrangements fall legally under the wide definition of 'settlement' which appears in Section 403 of the Income Tax Act, 1952: ' "settlement" includes any disposition, trust, covenant, agreement, arrangement or transfer of assets.'

Family settlements have a long history among the landed aristocracy of England. They reflected the fundamental importance of landownership in the nineteenth century, and they often embraced a concept of kinship extending far beyond the limited nuclear family of husband, wife and children. Although the subject has not attracted the attention of sociologists and anthropologists it would seem, from the historical studies that have been made, that the reasons for family settlements in the past were somewhat different from those which inspire similar arrangements today.[1] While basically concerned with the preservation and maintenance of great estates, they were seen more as a protection against the personal extravagance of members of the kin than as a means of avoiding taxation. There were, of course, many other purposes in these contracts. Settlements permitted life tenants to borrow extensively, often from insurance companies who appear to have played an important role in the evolution of certain types of settlement. As we shall see later, they still do so today though in a very different context.

We are not here concerned with the historical development of these instruments of social and economic power. It is sufficient to point out that their origins and growth in a particular class structure explain in some measure the uniquely generous provision now made in the United Kingdom for the alienation of income. The methods employed and the purposes underlying the use of these instruments have changed significantly in recent decades; moreover, those who now shelter under them extend well beyond the great landed classes.

Yet it is rare to find any recognition of these developments in the literature of public finance and income distribution. Legal writers have pointed out, for example, that discretionary trusts, carefully drafted to comply with the Rule against Perpetuities, can effectually preserve 'family' capital-income for up to a hundred years. Some substantial part of what is preserved, split and spread over the kin group, thus disappears for decades from the traditional pigeon-holes of the Board of Inland Revenue. To the irreverently minded lawyer, these and other devices are known as 'the disappearing trick'. A sizable number of these tricks could, by accumulating over the years, make nonsense of one end of the income statistician's Lorenz curve.

III

The Royal Commission on the Taxation of Profits and Income considered in 1955 some of the taxation issues presented by covenants and trusts. Its approach was not to ask whether there should be any

---

[1] Two recent studies which throw some light on the role played by family settlements in the nineteenth century are: Thompson F. M. L., 'The End of a Great Estate', *Economic History Review*, 1955, Vol. VIII, No. 1, p. 36, and Spring D., 'English Landownership in the Nineteenth Century', *Economic History Review*, 1957, Vol. IX, No. 3, p. 472.

general lightening in the existing limitations on the use of covenants but whether there was a case for imposing restrictions. At the Commission's request, the Board of Inland Revenue surveyed the problem but thought it unnecessary 'at the present time to propose any further conditions'.[1] No detailed facts were published by the Commission in regard to the number and growth of settlements as a whole, but it noted that the use of discretionary trusts was increasing. It also reported that, according to estimates made by the Board, about 75,000 deeds of covenant in favour of individuals were then current, and that tax forgone amounted to some £12,500,000 annually of which £7,500,000 represented surtax.[2] It is not clear what particular types of settlement these estimates covered, nor do we know how many beneficiaries were involved and to what extent. More recent estimates, showing a considerable growth in tax forgone, are quoted later.

Notwithstanding the views of the Board, both the Majority and Minority Reports made a number of recommendations designed to limit the use of covenants for tax avoidance purposes, the most far-reaching of which was to the effect that covenants made in favour of discretionary trusts should no longer be recognized for tax purposes. No action has been taken by the Government to implement this recommendation although other proposals (in these cases favourable to surtax payers) have been acted upon since the Report was published in 1955.

The whole problem of drawing a distinction between gifts on the one hand, and compensation on the other, lies in the fact that the British system of taxation is almost unique in the world in recognizing a payment voluntarily undertaken by the taxpayer as a charge on his income, provided that payment (a) is made under a promise backed by a deed of covenant extending to a period of more than six years; (b) is not made in favour of an unmarried infant of the covenantor;[3] (c) is not in exchange for value received from (or goods or services rendered by) the recipient.[4] In general terms, therefore, the intention of the law is to treat voluntary gifts and contributions, provided they are intended to be continued for longer periods, as charges and not as applications of the income of the taxpayer. Obviously, this conception of the law allows—and indeed encourages —real exchanges or payments to be dressed up as gifts, and makes

---

[1] *Final Report*, Cmd. 9474, 1955, p. 52.

[2] *Ibid.*, p. 51.

[3] Before April 22, 1936, covenants in favour of unmarried infants were effective for tax purposes in addition to (as now) married infants.

[4] Covenants by men in favour of their mistresses can only be regarded as ineligible for tax relief if the Inland Revenue can establish that they are given 'for services rendered' (see Monroe J. G., 'Annual and Other Periodical Payments', *British Tax Review*, December 1956, p. 289).

possible (as the Royal Commission pointed out) 'private understandings by virtue of which the benefit of the income never really leaves,
or is somehow returned to, the covenantor'. These understandings,
said the Commission, do exist; they are often described as 'gentlemen's agreements'; and 'they are no better than a fraud on the
system'. 'It seems singularly inappropriate,' concluded the Commission, 'to find that this phrase is sometimes invoked to describe a
family understanding which both parties fully intend to give effect
to but which, to attain a tax advantage to which they are not properly
entitled, they succeed in persuading themselves that they regard as
"merely moral".'[1] As we shall see, the extent, nature and consequences of these 'gentlemanly frauds' now current are not known.

The terms of reference of the Commission, though not extending
to include the matter of property transfers and death duties, were too
broad to allow it to consider in any detail all the ramifications of
covenants and settlements as a means of income and surtax reduction. These are set out, though mainly in legal form, in such studies as
Professor Wheatcroft's *The Taxation of Gifts and Settlements*, *Tax
Planning with Precedents* by Messrs D. C. Potter and H. H. Monroe,
and *Tax Planning and the Family Company* by Mr D. R. Stanford.
We have drawn liberally from these and other studies in summarizing some of the features of the law relating to covenants and
settlements. As we have already said, the selection of material has
been directed towards understanding the effects of these developments on the income distribution statistics. We thus make no attempt
to assess the extent of tax avoidance or to impute any unworthy
motives among those who make use of lawful procedures.

In taxation law, the word 'gift' normally denotes a voluntary
transfer of property from one person (the donor) to another (the
donee). There may be several donors or donees; they may be individuals or corporations, and in taxation law the word is sufficiently
wide to cover cases where the property is transferred to trustees on
trust for a number of beneficiaries or retained by the donor subject
to an effective trust in favour of others. As Professor Wheatcroft
explains, gifts may be made indirectly without any actual transfer of
property from donor to donee. They may be effected by creating a
burden on property of the donor or by releasing some right to which
he was entitled over property of the donee. A disclaimer of a legacy
may be a gift to the residuary legatee. Paying a debt which had
previously been voluntarily incurred is treated as a gift, as may also
be a transfer of property for value when the consideration was
originally derived from the deceased and the transfer is made in
satisfaction of an antecedent liability. 'Many cases occur,' wrote
Professor Wheatcroft in 1958, 'in connection with company or

[1] *Final Report*, Cmd. 9474, 1955, p. 53.

partnership transactions when one shareholder or partner benefits at the expense of another without there being any direct transfer from one to another.'[1]

As a means of avoiding income tax and particularly estate duty, outright gifts are, it is said, to be preferred in many circumstances to discretionary trusts.[2] Despite the Finance Act 1957, which applied new rules for estate duty purposes to all property comprised in a gift *inter vivos*, there 'would appear to be as many—although different—avoidance devices open under the new rules as there were under the old'.[3] Since this was written the Government has substantially liberalized the estate duty liability on gifts.[4] Instead of a five-year period[5] (when gifts were dutiable), the Finance Act 1960 provided for a reduction of 15 per cent in the principal value of the property if the death of the donor takes place in the third year; in the fourth year 30 per cent, and in the fifth year 60 per cent.[6] Only a year before this Act, the Economic Secretary to the Treasury had refused to change the five-year rule on the grounds that 'it was by no means excessive, especially as we, unlike many countries, have no gifts tax'.[7] These new provisions, under which income taxation as well as estate duty can be effectively reduced, are likely to make gifts even more popular, particularly when account is taken of the improved expectation of life among the wealthier classes. The decline in mortality rates since 1946, when the five-year rule was established, must have meant that, other things being equal, fewer gifts and settlements have been caught for estate duty liability. This factor does not seem to have been taken into account by the Government in lightening the burden of death duties in 1960.

The remarkable rise in the number of 'top-hat' and other pension schemes in recent years, under which tax-free lump sums of up to £40,000 or more may be paid on retirement, have an obvious bearing on this matter of gifts *inter vivos*. With large lump sums available, an annuity for life may be purchased and the remainder of an individual's

---

[1] Wheatcroft G. S. A., *The Taxation of Gifts and Settlements*, 1958, p. 111. For examples concerning shares in family private limited companies see *The Complete Death Duty Manual*, obtainable from Hogg, Robinson and Capel-Cure (Life and Pensions) Ltd, EC3. Detailed calculations are there given on how to save up to £45,000 duty on an estate of £76,000 and how, at the same time, to increase net income through reductions of income and surtax (1960 rates).

[2] Wheatcroft G. S. A., *op. cit.*, p. 73.

[3] *Ibid.*, p. 62.

[4] A reform urged upon the Government by such professional bodies as the Association of Certified and Corporate Accountants (*The Times*, January 23, 1959).

[5] The exemption period is only one year in the case of gifts for public or charitable purposes.

[6] Finance Act 1960, S. 64 (1).

[7] *Hansard*, H. of C., June 15, 1959, Vol. 607, Col. 83.

capital handed over to his heirs. His liability to tax will be heavily reduced and no estate duty will be payable if he survives for five years. The expectation of life of men at age 65 in many occupations in Social Class 1 must now be 15 years or more; of women three years longer at the same age.[1] Adjusted 'upper class' mortality rates (average of Social Classes 1 and 2) in 1951 showed, for example, that 'upper class' men aged 65–74 had an advantage of $12\frac{1}{2}$ per cent over the general population.[2]

No statistics have ever been published concerning the extent to which substantial fortunes are wholly or partly 'given away' at age 60–65 (or any other age). 'But most lawyers in large practices will confirm that this method of avoiding estate duty is extensively used.'[3] If this is so, the incomes of rich men and women will be reduced each year as more reach retiring age, thus affecting the official distribution of income statistics. According to the Chancellor of the Exchequer, an average of £6,000,000 per year was collected from 1946–7 to 1958–9 on gifts *inter vivos* which failed to escape death duty.[4] The total amount of gifts which did escape may have been many times larger. It is, wrote Professor Wheatcroft in 1958, 'probably the most common type of transaction today which is entered into for the purpose of avoiding or reducing tax liability'.[5] *The Economist*, puzzled by the apparent relative fall in the investment income of the rich during the 1950's as compared with the 1930's, came to the conclusion that an important factor was that the rich were passing on a lot of their property *inter vivos* to their children.[6]

One substantial exception to the revised five-year rule must be noted. Gifts in consideration of other people's marriages are exempt. No estate duty and no *ad valorem* stamp duty at all are payable. There is, therefore, in fiscal law a clear encouragement to marriage among the wealthier classes—and especially to early marriage. A

[1] For all men, England and Wales, it was 12 in 1957–9 (15 for women). *General Register Office, Quarterly Review*, 1960, No. 446, App. B.

[2] England and Wales (Lydall H. F. and Tipping D. G., *op. cit.*, p. 100).

[3] Wheatcroft G. S. A., *The National Tax Journal*, 1957, Vol. X, No. 1, p. 51. 'Estate duty saving,' wrote Mr Campion in 1959, 'has been a thriving industry for many years.' (Campion D. J. M., *British Tax Review*, November-December 1959, p. 406.)

[4] *Hansard*, H. of C., February 16, 1960, Vol. 617, Col. 105.

[5] He gave an example to show how important the saving was in 1958. 'If A has assets worth £120,000 and dies, he pays 50 per cent (£60,000) in duty. But if more than five years before he had given £60,000 to his children and maintained himself at his old rate of net income by selling capital each year, he would only leave £55,000 or less on his death. The appropriate rate on £55,000 is 35 per cent, so that duty is under £20,000. The beneficiaries save over £40,000 in duty, and in addition have the income of the £60,000 from the date of gift, which compensates them for the expenditure out of capital by A during the same period.' (*The Taxation of Gifts and Settlements*, p. 188.)

[6] *The Economist*, January 14, 1961, p. 112.

million pounds could, for example, be given to a son or daughter on marriage (under or over age 21) and no estate duty will be paid even if the donor dies the next day.[1] 'A marriage in the family,' wrote Mr Milton Grundy, 'may present a convenient opportunity for estate duty saving.'[2] Indeed, when discretionary trusts are associated with marriage settlements (as in the case of the Courtauld settlement of £450,000 in 1956) the saving of estate duty may be quite remarkable apart from the effects on income taxation. In this particular case, it was held in 1961 that on the death of the settlor within the statutory five-year period no duty was payable in regard to property trans-ferred to the trustees of a settlement expressed to be made in con-sideration of the marriage of the settlor's daughter, which took effect only on the marriage, even though, by the terms of the settlement, the property could, at the discretion of the trustees, be applied for the benefit of a very wide class of persons, many of whom were not within the marriage consideration.[3] This decision could clearly lead to a faster growth in discretionary trusts with far-reaching effects on estate duty revenue.

Though, in general, covenants may not be made in favour of one's own infant children, an exception is allowed for infants married under the age of 21. Income accumulated for such infants (or other people's infants) to which they become contingently entitled on attaining a certain age, is not assessable to surtax. When this age is reached, the accumulations arrive as capital and are thus never assessed for surtax. In such ways, 'family' liability to income tax, surtax and estate duty may all be heavily reduced. The Board of Inland Revenue has no information as to the extent or amounts currently involved in accumulation trusts of this kind.

IV

The making of outright gifts to members of the kinship, to other persons (including employees), to charity or for public purposes is only one of the ways by which an individual may reduce his liability for income tax, surtax and estate duty. All the law books and articles on tax planning in recent years agree that the most popular avoidance method today and the most common type of settlement is the deed of covenant by which a person, the covenantor, binds himself to pay a sum or sums of money in the future. A covenant may provide for a

---

[1] The Chancellor of the Exchequer was asked in May 1960 how much death duty was lost on account of gifts *inter vivos* made in consideration of marriage. His answer was that no information was available (*Hansard*, H. of C., May 20, 1960, Vol. 623, Col. 154).

[2] Grundy Milton, *Tax Problems of the Family Company*, 1956, p. 7.

[3] *Rennell v. IRC* (1961), *The Times* Law Report, October 31, 1961.

capital payment or payments or for a series of income payments. According to Professor Wheatcroft, the latter 'are far more numerous and offer much greater possibilities for reduction of taxation'.[1]

In a number of articles in the *British Tax Review* in 1956 on 'Modern Methods of Minimizing Taxation', Mr D. C. Potter, writing on the use of seven-year covenants, came to the conclusion that 'there is undoubtedly a serious loss of revenue to the Crown by reason of the use of seven-year covenants as a means of tax evasion, by which is meant the use of dishonest or fraudulent methods of avoiding the payment of tax properly payable'.[2]

After this comment, Mr Potter proceeded to show what legally can be done by a seven-year covenant. Such covenants can be made in favour of a wife not living with her husband if they are separated under a court order or a deed of separation. But to obtain a deed may hinder the chances of eventually getting a divorce. In such cases, where the parties are 'in fact separated in such circumstances that the separation is likely to be permanent' they are assessed as two, not one, persons, and so a seven-year covenant by the husband may provide considerable tax relief.

Substantial surtax payers living apart from their wives—or from one or more ex-wives—can, therefore, if the wife has little other income, make arrangements so that the wife or ex-wives are largely maintained after separation or divorce at the expense of the community. Tax relief may also be obtained in respect of settlements in favour of any children of such 'broken' marriages.[3]

'The odd situation had now come on us,' commented Lord Justice Hodson, 'in which a man of large means who paid surtax was able to have several wives . . . and to be very little worse off . . . whereas a working man who paid neither surtax nor any considerable amount of income tax found it quite impossible to comply with the law which still in this country enabled divorced wives to obtain maintenance from their husbands and also to provide for an excess of wives.'[4]

These 'social security' provisions in fiscal law for the rich may be extensively used today because of the large number of 'broken marriages' among members of Social Class 1 in the Registrars-General statistics. Income distribution data for single and married income units in the surtax class are likely, therefore, to be more significantly affected today than in 1938. It has, indeed, been estimated that the subsidy from taxpayers on account of separation

---

[1] Wheatcroft G. S. A., *op. cit.* (1958), p. 125.

[2] Potter D. C., *British Tax Review*, June 1956, p. 38.

[3] The settlement by the Marquess of Northampton in 1946 (divorced from the Marchioness in 1958) involving a trust fund of £227,000 in 1960 illustrates some aspects of this matter. It was reported in *The Times* on February 2, 1960.

[4] Lord Justice Hodson reported in *The Times*, July 31, 1957.

agreements and divorce orders among surtax payers amounted to the remarkable figure of £10,000,000 a year in 1959.[1]

Apart from these exceptions, a husband cannot, as a general rule, make any saving of tax by covenanting with his wife; nevertheless, there are various ways of minimizing income tax and estate duty by covenants in favour of children. They may be made for married infants, adult children whether married or not, in consideration of marriage, and for unmarried infants in any year of assessment in which the parent is not chargeable to tax as a resident of the United Kingdom. Extensive saving of income and surtax and estate duty may also be effected by settlements for the benefit of adult children and their children, born and unborn. The amount of tax and estate duty currently forgone by the present generation of taxpayers on account of covenants by the wealthier classes in favour of children yet unborn is not known, and doubtless could not be calculated. Such arrangements often form part of a family settlement. 'If the property vested in the head of the family,' wrote Messrs Potter and Monroe, 'is regarded as the property of the family, it is a *normal* (our italics) step for him to spread his property over the whole family during his life, and if this is done the tax and duty imposed upon the property will thereby be reduced.'[2]

One consequence of all these methods of spreading income-wealth may be that young people on attaining age 21 come into possession of substantial amounts of capital. The Oxford Survey in 1956 of the 'Savings and Finances of the Upper Income Classes' showed that the mean gross income of those aged 18–24 was £1,690.[3] Mr Lydall's study of *British Incomes and Savings* showed that 1 per cent of household heads aged 18–34 owned stocks and shares valued at £2,001–£10,000.[4] These studies, and the frequency with which those sampled in the former survey reported income from trusts, settlements and inheritances, provide some evidence of the prevalence of methods of spreading and splitting family worth.[5] The value of these methods resides largely in the fact that, in British fiscal law, a child's income is not aggregated with that of his parents.

[1] Wheatcroft G. S. A., *British Tax Review*, November-December 1959, p. 395.
[2] Potter D. C. and Monroe H. H., *Tax Planning*, 1959, p. 71.
[3] Klein L. R., Straw K. H. and Vandome P., *Oxford Institute of Statistics Bulletin*, 1956, Vol. 18, No. 4, p. 315.
[4] Lydall H. F., *op. cit.* (1955), p. 95.
[5] For example: 'With the group of recent heirs in our sample lists it may seem that we have an abnormal preponderance of beneficiaries of bequests. Some tabulations of the other five sample groups are striking in the frequency and amount of bequests received. At some time or other in their lifetime' (the average age was about 50) '52 per cent of these other groups had received an inheritance. For the people sampled from the *Directory of Directors* and *Debrett*, there was a mean value of inheritance, exclusive of house or land property, of about £500 during the past twelve months. Over their lifetime, such people had a mean

A more recent study by Mr Lydall and Mr Tipping of the distribution of capital between age groups in Britain in 1954 strongly supports this thesis. They found, from their analysis of estate duty statistics, that whilst average wealth increased with age, the greatest degree of wealth concentration is to be found in the age group 20–24. 'Of those aged 20–24 less than 2 per cent have more than £2,000; and the dispersion amongst these is very wide. Average net capital held by those with more than £2,000 varies from nearly £16,000 per person for those aged 20–24 to less than £10,000 per person for those in the age groups 45–74. In the final age group (75 and over) it rises to about £11,000. Amongst those with over £100,000 the average for the under 45's in 1951–6 was as high as £450,000, compared with less than £250,000 for those above this age.'[1] An analysis by sex in addition to age might well have found an even greater concentration of wealth among young men.

In 1959, Mr E. B. Nortcliffe conveniently summarized for readers of *Progress* some of the advantages of income covenants. 'Men maintain their former wives, mothers-in-law, adult sons taking postgraduate courses at university, invalid nieces, infirm uncles, wards, protegés and others who for undisclosed reasons have claims on their generosity. The technique used ensures that when a part of a man's income is transferred to someone else, the liability to tax on that part is transferred with it. Thus a married man with an income of £15,350 a year can put fully £250 in the hands of his aged and impoverished mother-in-law at a personal cost to himself of £28 2s 6d.'[2]

While marriage settlements (or marriage gifts) may be less widely used than outright gifts or family settlements on infant children they still offer, according to various tax planning authorities, many substantial advantages. They derive from the 1910 Finance Act which exempted 'gifts in consideration of marriage' from what may now be called 'the revised 1960 five-year rule'. They appear to have been designed to encourage marriage among the wealthier classes in Edwardian times—if not to strengthen family and kinship ties. No such encouragement for other classes is apparent in National Insurance law; indeed, in many respects, there exist positive disincentives.

It seems that gifts under marriage settlements may be made not only by near relatives (parents, grandparents, aunts, uncles and so forth) but by any person 'who has a genuine interest in seeing the

---

inheritance, exclusive of house and land property, of more than £5,000. The amounts inherited by those who actually had a bequest in these two groups was much higher, at about £10,000 over their lifetime.' (Klein L. R. *et. al.*, *op. cit.*, pp. 315–6.)

[1] Lydall H. F. and Tipping D. G., *op. cit.*, p. 96.
[2] Nortcliffe E. B., *Progress*, Unilever, Spring 1959, p. 76.

marriage established'.[1] Such persons may, therefore, include employers and personal friends. These settlements may confer benefits on persons other than the husband, wife and unborn children; for example, on some collateral relative in the event of the husband and wife failing to have issue. The ramifications of income tax, surtax and estate duty saving to all the parties concerned, both immediately and in the long run, cannot be pursued in this study. And, here again, the cost of these marriage and kinship benefits borne by the generality of taxpayers is not known. Quite apart from other considerations, the cost may well be higher today in consequence of the rise in marriage and remarriage rates (and the increase in size of family) exhibited by Social Class 1 occupations since 1938.

v

Fiscal and Other Benefits for Children

To treat this subject in any comprehensive way demands a full-length book in itself. Any review of this scope would have to survey all social security and dependency benefits, subsidies, direct and indirect grants for infant children, married children, grandchildren and unborn children through such media as child allowances, family allowances, gifts *inter vivos*, marriage and family settlements, covenants and discretionary trusts, educational covenants, endowment policies, employers' educational trusts and grants, university awards, professional training provisions, charitable reliefs, rate relief for public schools, and benefits derived from appointing infants as directors of controlled companies.[2]

With such variety and liberality of provisions for the offspring of the wealthier classes it is not surprising that the public and preparatory schools should be faced with the longest waiting lists in their history. But the demands for expansion today may be only a hint of what may come in the future if present fertility trends continue and the four-child family becomes a 'social norm' for the upper classes. This problem is not our concern. The continuance of 'family endowment' on the present scale by the State, in combination with rising fertility, does, however, raise questions concerning the effects on surtax and estate duty revenue. Such developments could lead to a further and substantial erosion of revenue which, along with all the other factors discussed in this study, might well threaten the structure

[1] Potter D. C. and Monroe H. H., *op. cit.*, p. 208.

[2] It is not known how many infants are directors (the Companies Act 1948 places no restriction on such appointments). In *Copeman v. Flood* (1941) 1 KB 202, an infant was cited with a remuneration of £2,600. This was not disallowed by the Commissioners. See also the reference to the organization of private companies with children as major shareholders in 'Savings and Finances of the Upper Income Classes' (Klein L. R. *et. al.*, *op. cit.*, p. 314).

of progressive taxation. What the effects are today in tax forgone and on current income distribution statistics it is impossible to estimate. In part, our ignorance derives from one of the striking features of British fiscal law which fails to aggregate the unearned incomes of children with the earnings of their parents. Such families are not regarded as forming common spending units. Social security law, on the other hand, holds firmly to the principle that parents and children form part of a single family and share the same standard of living. As Mr Prest has observed, the majority of the Royal Commission on Taxation 'almost tied themselves in knots arguing against this proposition'.[1] They did so partly on the grounds that there was no evidence 'before the Commission' as to the scale upon which gifts, settlements, trusts, 'charities' and other devices were putting children into possession of substantial incomes during the lives of their parents.[2] As no attempt was apparently made to search actively for evidence this statement by a Royal Commission appears singularly inept.

Nevertheless, we can give some general indication of the potential importance of discretionary trusts, money covenants and other tax-saving methods in changing the picture we have of the size distribution of incomes. In law, as we have already pointed out, the child is a separate tax entity from its parents and can therefore claim its own personal, small income and reduced rate reliefs against its own income. There is also a special provision which enables a child to recover back tax deducted from income from a fund which has been accumulated for him under a trust to which he was only contingently entitled on attaining a specified age or marrying. If the child has not already been allowed all his personal and reduced rate reliefs against income to which he was actually entitled, he can retrospectively recover the balance against contingent income when the contingency occurs.

Although income derived from a parent cannot be utilized for recovery of the personal and other tax reliefs of an unmarried infant —except in the case of a contingent accumulation settlement—never-the-less, so long as there is no reciprocal arrangement and the property settled is not provided by the parent, any other relations can make a deed of covenant or property settlement in favour of a child. If a deed of covenant is used then the income must be paid or applied within the tax year for the child's benefit; if a settlement of property is executed then the income can also be accumulated. The child's reliefs can then be recovered from the Revenue to the extent of the income and if the relative is a surtax payer he saves on this as well.

[1] Prest A. R., *Public Finance*, 1960, p. 274.
[2] *Second Report*, Cmd. 9105, 1954, p. 38.

F

Some references have already been made to the income and capital spreading effects of gifts and settlements in favour of children. As regards the immediate saving in tax, Professor Wheatcroft gives an example in his book of £80 a year provided by a grandparent (whose income is over £3,000 a year) for a child who has no income. In 1958, this would have cost the grandparent only £30; 'the remaining £50 will come from the Revenue'.[1]

This is an illustration of what Professor Wheatcroft considers to be the 'most common type of settlement today'; namely, the deed of covenant by which the covenantor binds himself to pay a series of money payments in the future.[2]

How widespread are these income covenants? In the Board's Memorandum of Evidence to the Royal Commission it was 'roughly estimated' that about 110,000 deeds in favour of individuals were current.[3] This estimate was derived from 'an examination of about 250 new deeds seen in a specimen period of seven days'—not, it may be said, a very extensive inquiry. It appeared that the average amount per deed was around £250. 'Assuming,' said the Board, 'that all the deeds were executed by taxpayers liable at the standard rate in favour of beneficiaries with no other income the loss to the Exchequer on account of income tax would be about £10,000,000. This can be regarded as an upper limit and the actual amount might be as low as £5,000,000.'

As regards surtax, the Board went back to an investigation in 1951 into the effects of decentralizing the assessment and collection of surtax. 'No special inquiry was thus made. From the figures then obtained it seems probable that some 40–50,000 of the deeds of covenant now operating are by surtax payers and that in the higher income ranges (£10,000 and over) about one-half of all surtax payers claim relief in respect of these deeds. The 1951 figures related to numbers only but it can be assumed that the average amount per deed for surtax payers is higher than the overall average—say £400. On these assumptions it seems likely that some £20,000,000 of income is involved for surtax and that the yield of that tax is reduced by over £5,000,000.'

When the Royal Commission published its *Final Report* in 1955 it quoted different figures. After saying that it would be 'a mistake to exaggerate the magnitude of the revenue at stake', it went on:

[1] Wheatcroft G. S. A., *op. cit.* (1958), p. 165. Covenants may be drafted by reference to the amount of the statutory limit of a child's income qualifying for child allowance, from year to year, in order to ensure that the maximum saving of tax is made even if the conditions for obtaining child allowance happen to be altered.

[2] *Ibid.*, p. 124–5.

[3] Board's Memorandum 119, *Vols. of Evidence, Royal Commission on the Taxation of Profits and Income*, 1952–5.

'From figures which the Board supplied to us at our request it appears that the total number of deeds of covenant in favour of individuals which are now current is about 75,000. According to the Board's estimates, the recognition of these covenants for tax purposes reduces the annual yield of surtax by some £7,500,000 and the annual yield of income tax by some £5,000,000.'[1] No attempt was made to assess the validity of these estimates—if they may be so described.

In answer to a question in February 1961, the Financial Secretary to the Treasury said that the amount of income tax and surtax forgone in 1954–5 in respect of covenants in favour of individuals amounted to £5,000,000.[2]

While recognizing that the extraction of such estimates must present great difficulties it is, nevertheless, hard to reconcile these various figures for the period when the Royal Commission was at work. Later estimates by the Board indicate a sharp and continuing rise.

In 1958 it was reported that the total sums payable under deeds of covenant amounted to £37,000,000 in 1956–7; of this sum £25,000,000 represented non-charitable deeds in favour of persons.[3] The income tax and surtax forgone was estimated at £6,500,000 and £10,000,000 respectively—a total of £16,500,000. Despite reductions in the effective rates of taxation since the Board submitted its Memorandum to the Commission, the surtax loss appears to have doubled.

In May 1960 it was estimated that the sums payable under deeds of covenant other than to charities had risen since 1956–7 by £5,000,000 to 'the order of £30,000,000'. The loss of tax was put at £15–20,000,000 'of which rather more than a third was of income tax and the rest of surtax'.[4] If the surtax loss for 1959–60 is put at £12,000,000 this would represent about 7 per cent of the total yield of surtax in that year. But all this is guesswork as no basis exists on which to check the Board's figures. No estimates have ever been published as to the effects on estate duty, and no consideration appears to have been given to the need for a census of settlements and trusts. The Royal Commission failed to recommend one.

The great and growing popularity of income covenants for a variety of purposes is due to the simple fact that, provided they comply with a number of not very onerous conditions, they effectively transfer income from one person to another. The covenantor pays the periodical sums due less tax, whilst the covenantee who receives his payment after tax has been deducted can claim from the Revenue a repayment of any tax so deducted to which his own tax

[1] Cmd. 9474, 1955, p. 51.
[2] *Hansard*, H. of C., February 17, 1961, Vol. 634, Col. 189.
[3] *Hansard*, H. of C., March 12, 1958, Vol. 584, Col. 62.
[4] *Hansard*, H. of C., May 19, 1960, Vol. 623, Col. 152.

exemptions and reliefs entitle him. For surtax purposes, the gross income under the covenant is deducted from the covenantor's assessment and added to the covenantee's.

It is clear, therefore, that when the covenantor is adult and relatively rich and the covenantee is young and nominally poor, large sums of money can be obtained from the generality of taxpayers.[1] In an extreme case (cited by Professor Wheatcroft in 1958) where the covenantor paid surtax at the highest rate and the recipient had no other income, the covenantor could provide £150 a year at a cost of £11 5s to himself, the balance of £138 15s being borne by other payers of taxes, duties and insurance contributions.

Another example from the tax planning literature relates to a covenantor with a total unearned income of £10,000 gross a year in 1959 who paid by deed of covenant £10 a week to his son aged 21 who had no other income except a university scholarship.[2] Of the total annual covenanted sum paid over of £520 the net cost to the covenantor was £104. The total received by the son (net under the deed plus repayment of tax) was £418 10s (scholarship income being exempt). In short, as a result of this family transaction, the generality of taxpayers found at 1958–9 tax rates an additional £314 10s a year in support of a young man already being educated at the public expense.

This particular case is but one example of the widespread use of deeds of covenant, discretionary trusts, assurance policies and other devices for the payment of education costs and the maintenance of children. Their use on the present scale must materially affect, in different ways, the Board's income distribution tables both before and after tax. According to the circumstances of the parents, the child, the grandparent, uncle, aunt, godparent, friend or other person(s) acting as covenantor or trustee(s), the saving of tax in 1960–1 may have been of the order of 50–60 per cent on sums of £400 or so a year. The School Fees Insurance Agency Ltd which, like many other similar concerns, has a flourishing business, advertises 'Public School fees at half price' in *The Times* and other 'quality' papers. This can only mean that the balance is found by the general

[1] This is, of course, only one illustration of what can be done by the use of covenants. Two incomes can, for instance, be averaged, thus saving on income and surtax. A case of this type was cited in the House of Commons in October 1958. A man and a woman, living together, with incomes of £5,000 and £1,500 respectively, agreed not to marry, the man then covenanting to pay £1,500 a year to the woman. Subsequently, twins were born and the couple then effected a discretionary trust which paid these two children £250 a year each less tax. The couple, who intended to marry when their children were twenty-one, estimated that they would by then have saved £20,000 in tax. This would be settled on the children (*Hansard*, H. of C., October 29, 1958, Vol. 594, Col. 239).

[2] Taken from Potter D. C. and Monroe H. H., *op. cit.*, pp. 7–8.

body of taxpayers. One leading tax consultant in London was reported in 1958 as saying that 'nine out of every ten covenants which he helps to draw up are concerned with education'.[1] A year later the Educational Correspondent of *The Observer* said they were 'increasingly popular with grandparents'.[2] These arrangements can, from the viewpoint of income distribution statistics, affect three tax units—the child, the parent and the grandparent. Life assurance relief for the child also comes into the picture, for policies may be drawn the premiums for which are paid for out of covenant money.

Sir Edward Gillett, acting for the trustees of Uppingham School in a rating assessment appeal, said that public school fees were often paid by covenant and by grandparents. 'That is being done today, to my knowledge, in a great many cases.'[3] Shrewsbury had 1,300 tax-free covenants current in 1956.[4] The Special Correspondent of *The Times*, reporting in 1959 on the fact that the unprecedented prosperity of preparatory schools was only equalled by that of the public schools, commented on the advantages of grandparents covenanting for their grandchildren, and friends for each other's children. These subsidies from taxpayers accounted in part for the very large expansion in the number of boys attending schools belonging to the Incorporated Association of Preparatory Schools and waiting lists 'longer than they have ever been'.[5] 'Subsidies', it may be objected, is an inappropriate word to use in the context of legally approved tax deductibles or reliefs. Nevertheless, it has been employed in this field by many authorities to describe various taxation arrangements. *The Economist*, for example, considers that any relief from or abolition of Schedule A tax would constitute a subsidy to houseowners.[6]

The question of a subsidy from the general population to particular groups or classes for particular purposes is also raised in the

[1] *Sunday Express*, September 28, 1958. They were described by Mr Bernard Harris as 'State subsidies'.

[2] *The Observer*, August 30, 1959.

[3] Reported in *The Times*, October 29, 1957.

[4] Reported in *The Evening Standard*, February 20, 1956. This report also described how three or more unrelated parents can make a chain of education covenants.

[5] *The Times*, July 21, 1959.

[6] *The Economist*, August 9, 1958, and April 4, 1959. The Financial Secretary to the Treasury described certain proposals for new tax concessions in 1956 as 'a kind of concealed subsidy to certain cinemas' (*The Times*, June 23, 1956). Sir Oscar Hobson regarded investment allowances as subsidies to industry (*News Chronicle*, January 21, 1959). In the USA, the term 'subsidy' appears to be generally applied to personal deductions—see US Dept. of Health, Education and Welfare, *Research and Statistics Note No. 4*, 1958; *Personal Deductions in the Federal Income Tax*, by Prof. C. H. Kahn, 1960, p. 144; and 'Rationale of the Medical Expense Deduction', by Prof. J. E. Jensen in *National Tax Journal*, 1954, Vol. VII, No. 3, p. 283.

field of rating relief. An Opposition amendment to rate public schools fully rather than give them a 'subsidy' in the form of 50 per cent relief as 'charities' was resisted by the Government when the Commons Standing Committee debated the Rating and Valuation Bill in 1961.[1] Similar relief to scientific societies and industrial research associations had earlier been described (and resisted) by the Minister of Housing and Local Government as a 'compulsory contribution' from ratepayers to these bodies.[2] Because of the operation of the Central Government's rate deficiency grant to local authorities it appears that this relief from the rates for public schools ultimately falls on the general body of taxpayers. This is certainly so in the case of Eton.[3]

Without a considerable amount of research it is impossible to estimate how much is involved in these rating subsidies or reliefs—as well as reductions in gross rateable values[4]—afforded to public schools. Any attempt to measure the full effects of education covenants, child allowances and other benefits for children would, however, have to take these matters into account.

Another popular and widespread method of saving tax and estate duty and paying for the education of someone else's child is to pay a once-for-all capital sum to the school, which then transfers the sum to a firm of insurance brokers for settlement of the fees as they fall due. The advantages of this method are that grandfather (or anyone else) may, if he pays eight years in advance, put down (as *The Economist* calculated in 1957) only £880 for fees that would otherwise total £1,380; and that the sum is not liable for death duties.[5]

Some illustrations from an earlier period of the historical role of insurance companies and agents in relation to education trusts are given in Appendix E. It is clear from this account that certain companies played an important part in devising schemes for 'fictitious trusts' and other practices condemned as unsavoury by the *Policy Holder*, one of the leading insurance journals.

Mutual covenants and other arrangements of a similar kind described above are not permissible by law, said Mr Macmillan, as Chancellor of the Exchequer in 1956. 'Mutual arrangements would be out of order.' The Board of Inland Revenue has, however, to

[1] *Hansard*, H. of C., Standing Committee D, February 16 and 21, 1961.

[2] *The Times*, February 8, 1961.

[3] Letter from the Clerk to the Council of Eton UDC, April 21, 1961.

[4] 'Noble hearts,' wrote *The Economist* in June 1960, 'will beat more easily as a result of this week's decision by the Lands Tribunal to reduce the rateable value of Shrewsbury School' (June 25, 1960, p. 1324). The arguments before the tribunal embraced the whole field of public school finance. For a summary of the case, which was important to many other public schools, see *The Times*, May 17 and 20 and June 21, 1960.

[5] *The Economist*, March 9, 1957.

establish that such arrangements between relatives and friends have, in fact, been made. As written agreements are rarely if ever concluded, the Board, it seems, has a very difficult duty to perform.[1]

A recent judgment by Mr Justice Danckwerts has a bearing on this matter. It is of sufficient importance to justify a lengthy quotation from *The Times* report.[2]

His Lordship said that this was a claim for certain personal allowances to be given to the respondent, a well known film actor, representing his children. The commissioners had decided that the respondent, on behalf of his children, was entitled to the allowances. The facts found by the commissioners were that in December, 1954, a company was formed under the name of Roehampton Productions Ltd, with a share capital of £100, divided into 100 shares of £1 each. Mr Hawkins was not at any time a shareholder in the company, although he was a director. In December, 1954, a service agreement was also made between Mr Hawkins and the company whereby he agreed to render to the company his services on the terms and conditions therein contained. By a settlement dated March 3, 1955, Mr Hawkins's father-in-law, Mr Horace George Beadle, settled the sum of £100 out of his own moneys for the benefit of his grandchildren and remoter issue. Mr Hawkins had not supplied Mr Beadle with any money for this purpose. The trustees used the funds of the settlement to subscribe for the 98 unissued shares in the company. The company received £25,000 for providing Mr Hawkins's services for a film called 'Fortune is a Woman', and paid Mr Hawkins £900. On October 18, 1956, the company declared and paid an interim dividend of £500, free of tax, which the trustees paid Mr Hawkins's children, subject to tax. That was the tax which was the subject of the claim for repayment. The commissioners found that Mr Hawkins was aware that steps were being taken to put into effect proposals of his accountants and solicitors, but that he was not consulted with regard to them and was not present at any meetings when the matter was discussed. The commissioners decided that there was no 'arrangement' to which section 397 of the Income Tax Act, 1952, could apply and that Mr Hawkins was not a 'settlor' within section 403 of the Act.

The settlor in regard to the settlement of March, 1955, was plainly Mr Beadle. It was contended on behalf of the Inland Revenue that the artificial meaning of 'settlement' in section 403 of the Act was to be applied to this settlement so that there was another settlor, namely Mr Hawkins, and he was to be treated as a settlor of the income fund which, therefore, was to be treated as

[1] *Hansard*, H. of C., April 10, 1956, Vol. 551, Col. 21. To call for declarations that no arrangements exist, as the Board does, is hardly likely to be completely effective (*Hansard*, H. of C., July 30, 1959, Vol. 610, Col. 148).

[2] *The Times*, November 15, 1960.

his. Taking the settlement of March, 1955, by itself it was plain that
Mr Hawkins was not in any way a settlor. It was upon the con-
struction of section 403 that it was sought to regard him as a
settlor. That could only be done by reading into the settlement the
facts which occurred on the formation of the company and the
service agreement. Mr Hawkins was a concurring party in the
issue of the shares in the company to the trustees, but only in his
capacity as a director, and the articles of the company required
that there should be two directors. He could not be said at any
time to have been in control of the company. At the date when the
company was formed and he entered into the service agreement,
the settlement had not been created and there could not, at that
date, have been any connexion between the transaction in regard
to the company and the settlement which was afterwards executed.
If it had been proved that a scheme had been made which involved
the formation of a company, the entering into a service agreement
by Mr Hawkins, a settlement of a sum of money, and the purchase
of the shares in the company with that money, it would be possible
to say there was an 'arrangement' within the meaning of section
403 of the Act.

The existence of any such arrangement had been negatived by
the findings of fact of the commissioners. Mr Hawkins was not a
party to any such scheme. That seemed to his Lordship to be the
material factor in the case. The sum of £25,000 was received by the
company as a result of acting by Mr Hawkins, but it was received
by the company and belonged to the company, and Mr Hawkins
was only entitled to £50 a week and his expenses. It had been
asked why Mr Hawkins should enter into any such agreement.
Having regard to the high incidence of surtax upon persons who
at one moment might make a high income but which income was
subject to vicissitudes it had become a common practice for actors
and actresses to enter into transactions of this kind for the very
natural object of trying to prevent their income being diminished
by surtax.

In his Lordship's view Mr Hawkins could not be made responsible for the provision of the dividend, for he was merely one of the
directors of the company. Nothing Mr Hawkins had done fell
within the terms of section 403 of the Act. On the facts found by
the commissioners, they reached the right conclusion of law and
his Lordship was not entitled to disturb it. The appeal would be
dismissed with costs.

Mr Justice Danckwerts' decision was reversed by the Court of
Appeal in May 1961. Leave to appeal to the House of Lords was
granted.[1]

This case is an illustration of the combined roles played by 'one-
man companies' (as they are called) and settlements in the avoidance

---

[1] *The Times*, May 4, 1961, and *British Tax Review*, May-June 1961, p. 819.

of income tax, surtax and estate duty.[1] From a survey of the tax planning literature it appears that it is not only actors and actresses who are instructed in these complex matters. It may well be that such methods of tax and estate duty saving are fairly common among a proportion of surtax payers, particularly the self-employed and certain members of the professional classes. It would thus follow that the Board's separate statistics on the incomes of such groups are hardly worth the paper on which they are printed.

Apart from other considerations, if the Board cannot establish the existence of these arrangements or ascertain their effects it obviously cannot 'feed back' into its distribution tables these transfers of income to their appropriate 'income units'. A wealthy taxpayer with a separated wife, a mistress and four children (all with 'separate' incomes under different schedules) might be represented in the Board's yearly tables by six or seven income units.

Nor is it easy to feed back, in terms of income at current values and to the appropriate unit, educational benefits in cash and in kind provided by or through the agency of employers. Much may depend on what they are called. 'Payment by a firm of the school fees of the children of a *selected* (our italics) number of its employees are normally reckoned as expenses wholly and exclusively laid out for the purposes of the firm's trade,' said the Financial Secretary to the Treasury in 1956, 'and accordingly are admissible as a deduction in computing the trading profits of the firm.'[2] Premiums paid by employers on endowment policies for the education of employees' children are similarly treated.[3] Fees or premiums are not regarded as income of either the employee or his children for tax purposes provided they relate to scholarships awarded on the basis of merit to children of employees earning less than £2,000 a year. It has also been held that where an employer covenants to pay an annual sum for seven years to be held on discretionary trusts for the benefit of a fixed number of named infant children of employees, a sum paid to a particular child must count as his personal income and not as a part of his father's income.[4] The employer concerned in this particular case was Imperial Chemical Industries Ltd.

As against a rise in salary, there are distinct advantages to both employers and employees in arrangements of this kind, quite apart from the effect of promoting the interests of particular 'public' and private schools. 'Industrial scholarships,' wrote Mr Lunt, 'though a

---

[1] Other relevant material may be found in 'Taxation and the Ownership of Real Property', Crump S. T., *British Tax Review*, March-April 1961, especially pp. 106–7. See also Chapter 6 of this study.

[2] *Hansard*, H. of C., May 3, 1956, Vol. 552, Cols. 55–6.

[3] *Hansard*, H. of C., June 3, 1960, Vol. 624, Col. 174.

[4] *Barclays Bank v. Naylor* (1960) 3 All ER 173; (1960) TR 203.

boon to the surtax payer, are open to some objections.' On behalf of the Incorporated Association of Head Masters, he criticized them because they took away some element of freedom of choice in respect of higher education, because they pre-empted 'promising but uncommitted human material', and because they are 'a sort of evasion'.[1]

Developments in these and allied fields, aided by the work of the Public Schools Appointment Bureau[2] and the growth of 'selection consultancy' organizations, are reflected in the changing social characteristics of top management in British industry. A number of recent studies, analysed in a report from the Department of Social Science at Liverpool University, show that there has been a considerable increase in the number of managers who have been to a public school, and that a high proportion of directors come from a relatively small number of such schools.[3] There would seem to be a connection between these trends and the growing use of covenants, discretionary trusts and other methods of splitting the income-capital of a family and meeting, in a variety of ways, the costs of private education.

Such transactions, widespread though they may be, will not, however, be discernible in the Board of Inland Revenue's income distribution tables. Nor will the activities of such bodies as the British Overseas Mining Association (recognized by the Board as a charity) which provides 'scholarships' at schools and universities for children of employees whose firms are members of the Association. Covenants are used and tax is recovered on the 'donations' made.[4]

In a letter to the Chancellor of the Exchequer in 1959, the Chairman of the Institute of Directors said that the fund for the advancement of scientific education in independent schools was 'almost wholly dependent on covenants'. The fund was then in debt to the extent of approximately £1,200,000; this sum having been borrowed against covenants.[5] Other educational schemes for tax-saving which make use of credit facilities, deduction of interest charges, insurance premiums, income payments treated as tax-free capital payments

[1] Letter to *The Times* from Mr R. G. Lunt, Chairman, Careers Committee, Incorporated Association of Head Masters, February 11, 1961.

[2] In 1956, the Director of the Bureau acknowledged the 'generosity of over 340 firms and companies' which had enabled the work of the Bureau to expand considerably over the last six years (Careers Supplement, *Liverpool Daily Post*, May 15, 1956).

[3] McGivering I., Matthews D. and Scott W. H., *Management in Britain*, 1960, esp. pp. 65–72.

[4] *The Times*, December 30, 1958. Similar schemes have been set up by the Institution of Civil Engineers, the Association of Consulting Engineers and the Federation of Civil Engineering Contractors (*The Times*, January 11, 1958).

[5] Letter from Major-General Sir Edward Spears reported in *The Times*, January 16, 1959.

and so forth, now promoted by professional associations, insurance companies, death duty consultants and hire purchase firms, are also not fully reflected as 'statutory income' in the income distribution statistics for parental tax units.[1]

How many of these arrangements for 'educational' purposes (in *lieu*, perhaps, of higher salaries) are recognized and classified as 'charitable deeds' is not known. The Board reported to the Royal Commission that the number of such deeds had risen from 294,000 in 1946 to nearly 500,000 in 1952.[2] In March 1958 there were said to be 800,000 deeds to charities.[3] Two years later the figure had risen (again in suspiciously round numbers) to 900,000 involving 'about £15,000,000 payments before deducting tax' (or another £3,000,000 on the 1958 estimate).[4] The tax loss on these deeds was put at £6,000,000 a year. In its *Final Report* in 1955, the Royal Commission, remarking on the increase in the number of annual claims under charitable deeds of all kinds, quoted an estimate by the Board of £35,000,000 representing tax forgone on account of the comprehensive exemption of charities from taxation (no estimate was given of the loss of estate duty).[5] This, said the Commission, amounted 'in effect to a grant of public moneys towards the furtherance of such causes as come within the legal category of charity without Parliamentary control of their individual purposes or of their administration'.[6] Two members of the Commission, Professor J. R. Hicks and Mr S. G. Gates, described the exemption in a Reservation as a 'blind and hidden subsidy contributed by the State'.[7]

There is, however, in revenue law no independent definition of a charity. The whole problem is surrounded by the mysteries of evolutionary analogies since the Elizabethan Statute of Charitable Uses, and many judges have declared themselves baffled by the task of deciding according to law what is and what is not a charity. Anyone can create and endow a charitable trust or transfer funds to an existing one. By a covenant in the appropriate form anyone can transfer a portion of his taxable income to any charity he selects. In 1952, the Nathan Committee estimated that on the average ten new

[1] See, for example, *The Times*, October 31, 1955, and July 21, 1959, and *The Manchester Guardian*, October 15, 1959. The relationship of interest charges to the financing of education is discussed in Chapter 6.

[2] Board's Memorandum 81, *Vols. of Evidence, Royal Commission on the Taxation of Profits and Income, 1952–5*.

[3] *Hansard*, H. of C., March 12, 1958, Vol. 584, Col. 62.

[4] *Hansard*, H. of C., May 19, 1960, Vol. 623, Col. 152.

[5] Estate duty is not payable on a gift for public or charitable purposes, provided the gift was made without the retention of a benefit more than one year before the death of the donor.

[6] Cmd. 9474, 1955, p. 55.

[7] *Ibid.*, p. 352.

charitable trusts came into existence every week.[1] There were then thought to be some 110,000 trusts in being. Professor Victor Morgan estimated the assets held by general charities in 1955 at more than £1,600,000,000.[2] Little is known, however, about the extent to which this vast sum is held on trust for genuinely charitable purposes in the sense of the relief of poverty and distress and the education of the indigent, or how far some of these trusts are being used as a means of conferring benefits in the form of education, social amenity, professional advancement, and other categories of welfare on those who are not poor and, also perhaps, as a means of avoiding estate duty. Over forty years ago, the Royal Commission on the Income Tax recommended that Parliament should specifically redefine 'charities' for the purposes of income tax.[3] Since then, the use of the term has come to embrace a far wider range of amenity and welfare for a far wider section of the population.

VI

Discretionary Trusts

We have referred on a number of occasions to the use of discretionary trusts, and have quoted the recommendation of the Royal Commission that such dispositions ought not to be allowed to rank for tax purposes. The Commission believed that the use of such trusts was increasing;[4] more recent evidence shows that they are now a popular means of avoiding income tax, surtax and estate duty.[5]

The advantages of this type of deed are considerable.[6] The pro-

[1] Cmd. 8710, para. 545 (f).

[2] Excluding the assets of churches, the universities and trade unions the total comes to £1,658,000,000 (Morgan E. V., *The Structure of Property Ownership in Great Britain*, 1960, pp. 127–41).

[3] *Report of the Royal Commission*, Cmd. 615, 1920, p. 68.

[4] Cmd. 9474, 1955, p. 52.

[5] See, for example, Wheatcroft G. S. A., *British Tax Review*, Nov.-Dec. 1954, p. 394, and *The Taxation of Gifts and Settlements*, 1958, pp. 184 and 217; Potter D. C. and Monroe H. H., *Tax Planning*, 1959, pp. 170–205; Cozens-Hardy Horne R., *British Tax Review*, 1957, p. 256 and Revell J. R. S., 'An Analysis of Personal Holders of Wealth', British Association for the Advancement of Science, 1960.

[6] They are set out in detail (with examples of the use of other methods) in the many brochures now obtainable from death duty consultants, insurance companies, insurance brokers and others. Considerable numbers specify the advantages of the acquisition of domicile outside the British Isles; the acquisition of immovable property outside the United Kingdom (e.g. certain types of mortgages in Jersey and property in Bermuda); conversion of the estate (or part of it) into agricultural property or standing timber; the acquisition of 'industrial hereditaments'; the creation of separate non-aggregable estates by means of a special type of insurance policy; the divestment of 49 per cent of the shares in a company in favour of wives or children in combination with trusts and life policies; and other methods. The advantages are not restricted to estate duty saving; certain methods also bring a considerable saving in income and surtax. Some of these advantages are also described in the next chapter.

cedure is for a covenant to be made with trustees constituted for the purpose, and the trustees are given the duty of distributing the annual payments that they receive among any one or more of a number of named beneficiaries and in such shares as they may decide. The trustees thus have full discretion to appropriate the money to one or more beneficiaries in one year and to others in another year and to vary the shares themselves from year to year. Sometimes the named beneficiaries are 'charities', sometimes named individuals; sometimes both classes are included. Clearly, one of the main attractions of a covenant in this form is that the covenantor is able to keep an effective voice in the destination of the income each year as between one beneficiary and another, assuming of course that the trustees whom he has chosen are prepared to consult him as to the exercise of their discretion. As one firm of death duty consultants has put it: 'although he (the covenantor) has removed certain of his assets from appropriation by the Chancellor, he has not passed them over absolutely to his successors to dispose of as they will, but only to trustees who presumably will have been selected because of the knowledge that they will co-operate with him in exercising their discretion in the way he would subsequently indicate.'[1]

The selection of trustees may thus be seen as one of the most important decisions in the life of the wealthy. As a power-supporting class, the trustees of such covenants represent another of the keys to understanding the sources and distribution of wealth in society. If chosen wisely, the covenantor not only retains the power that resides in the continuance of freedom of choice in the use of capital and income through the operation of the principle of uncertainty of interest, but he also secures a substantial reduction in taxation and estate duty. Power is thus enhanced in the attractive garb of benevolence.

It can also be preserved for a long time. As the law now stands, the settled property under certain trusts may be preserved for a century or more without the payment of one penny of estate duty. Theoretically, there is a limit of eighty or ninety years which is imposed by the Rule against Perpetuities; this requires the final limitation of a non-charitable trust to take effect within twenty-one years from the death of a 'life in being' when the deed is drawn. But as nearly all discretionary trusts enable the trustees to distribute the capital within the specified class, any intelligent trustees would certainly do so before the risk of estate duty arose.[2] After, so to speak, a second

---

[1] *The Complete Death Duty Manual*, 1960, obtainable from Hogg, Robinson and Capel-Cure (Life and Pensions) Ltd, EC3.

[2] Trusts may also be broken up during the lifetime of the life tenant under certain conditions. Provided the life tenant does not die within the revised five-year period, no duty is then payable on his death. In the valuation of trusts 'broken-up' and the sale of reversions there are also possibilities of tax avoidance.

'probationary' period of some risk for five years (a risk which is much lower today than in the past because of the reduction of mortality rates among Social Classes 1 and 2) the capital then enters on another duty-free career.

Life policies are frequently used today for gifts, settlements and discretionary trusts to insure against these risks. Lord Glenconner, Chairman of the Northern Assurance Group, mentioned in 1959 two perhaps noteworthy policies of this type, each for £1,650,000, and said that 'an increasing number of large proposals of this type are being received'.[1] Policies effected under the Married Women's Property Act, 1882, or to be held from their inception on trusts excluding the settlor, offer substantial estate duty advantages. As Professor Wheatcroft points out, 'Even though the settlor pays all the premiums, so long as he never had a beneficial interest in the policy, its proceeds on his death will be treated as a separate estate and only aggregated with other similar interests of the beneficiaries. In addition, once he has lived two years from the date of the policy, further relief is obtained on premiums paid more than two years before his death; by the fifth year only 79 per cent of the policy will be dutiable. After five years still further relief is obtained as the proportion of the policy represented by premiums paid more than five years before death is excluded altogether; the slice of the policy then dutiable is calculated by the formula $79/20n$ where n is the number of annual premiums which have been paid by the settlor.'[2] If, because of the state of the settlor's health, there is any difficulty in getting an assurance company to accept the risk, this problem can be solved by purchasing an annuity at the same time from the same company. Since the Finance Act, 1956, the tax advantages of a purchased annuity may be quite considerable to an elderly person as only the income content is taxable, whilst the capital content is tax free.

Not only are assurances of various kinds commonly used with settlements when it is desired to accumulate income (because of the doctrine that a trust to pay life policy premiums is not liable to be void as an accumulation) but, as all the tax planning books point out, they are favourably treated from a taxation aspect compared with other types of property. It cannot be said that the premium income from such policies is 'saving' in the sense commonly used by assurance companies.[3] There may be some element of such saving in policies taken out under the Married Women's Property Acts

---

[1] Company meeting report, *The Economist*, May 16, 1959, p. 688.

[2] Wheatcroft G. S. A., *British Tax Review*, November-December 1960, p. 400.

[3] The annual and meagre statistics published by the Life Offices' Association purport to show that all assurance premium income is the result of savings by *individuals* (our italics) 'to provide security for their families' (see 1959–60 Report). Other examples of the misleading nature of these statistics are given later.

whereby the proceeds arising upon the death of the husband (or wife) are free from aggregation and are not included in the estate for duty purposes. In most cases, however, the real object is to reduce estate duty and to obtain certain tax advantages.

These are made plain in the literature distributed by death duty consultants, assurance companies and brokers. As one Manual put it: 'insurance obviously plays a very large part in the solution of almost every problem' of estate duty and tax saving.[1] In a typical example, relating to a saving of £20,000–£30,000 for a man aged 50 with a wife, two children and an estate of £75,000, the method proposed was to establish a discretionary trust and to take out three whole life with profits policies written as three equal policies under the Married Women's Property Act.

Some glimpses of the history of avoidance between the early 1940's and 1954 in relation to schemes of this kind are given in Appendix E. They afford some idea of the scale of the problem and its development during the period when all income distribution studies were reporting an increasing trend towards equality. They also provide further evidence of the critically important role played by insurance in facilitating the rearrangement of income and wealth.[2]

Yet it is difficult, as Mr Lydall and Mr Tipping found in their study of personal wealth, to estimate the effects with any precision. They pointed out that, according to published data, the aggregate of sums assured under life policies early in 1954 was nearly £8,000,000,000.[3] 'But the estimate from the estate duty returns of the value of life insurance policies held by persons with more than £2,000 of net capital in 1954 amounts to less than £900,000,000. It is impossible to believe,' was their comment, 'that this is a correct measure of the total of sums assured by this group; and it seems that, for some reason, the value of life insurance policies recorded in the estate duty returns is understated.'

Another and even more serious gap in our knowledge about the distribution of income and wealth relates to discretionary trusts. Literally nothing is known about them in terms of numbers, beneficiaries and amounts of capital involved. The extent of our ignorance is abundantly clear from the questionnaire printed in Appendix A.

---

[1] The Complete Death Duty Manual, op. cit., 1960 edition, p. 15. See also booklet Estate Duty and Insurance and Notes on the Married Women's Property Act, 1882, distributed by the Norwich Union Life Insurance Society in 1961.

[2] This role has incidentally meant that the Duke of Argyll (to take one example) has 'no power of gift' in relation to the family (Campbell) heirlooms. According to a judgment by the Lord President of the Court of Session in Edinburgh, the only person entitled in 1953 to demand delivery of these heirlooms was the Equity and Law Life Assurance Society Ltd. (The Times, August 10, 1961.)

[3] Lydall H. F. and Tipping D. G., op. cit., pp. 102–3.

But even these questions, which cannot be answered, do not exhaust the subject; what is needed is a national census.

Because of the darkness that surrounds the whole subject most income statisticians have ignored these trusts. Mr Revell drew attention to their importance in his paper to the British Association in 1960 analysing the personal holders of wealth.[1] Stimulated by this paper, Mr Lydall and Mr Tipping made some allowance for them in their study of the distribution of personal wealth.[2] They assumed— on the basis of little more than guesswork as they emphasized—that in 1936 the top 10 per cent of property owners in Britain owned £300,000,000 of discretionary trust funds and that by 1954 the total had risen to £1,000,000,000. But even this figure—large as it is— could be a substantial understatement. Values could have risen more than threefold between 1936 and 1954; they have certainly increased very substantially since then with the rise in real property values; and all the evidence in this study and in the tax planning literature points to a considerable growth in the number of such trusts in existence.

This is one of the two major forms of settled property which, as Mr Revell says, 'completely escape death duties and which are sufficiently important seriously to upset estimates of total wealth in the beneficial ownership of individual persons'.[3] The other is property covered by the surviving spouse exemption.

The total wealth at present values of persons domiciled in Great Britain has been estimated, in very round figures, by Mr Revell at £50,000,000,000.[4] Of this, about £4,500,000,000 is held in trust for individual persons (settled property). An addition has also to be made for non-dutiable settled property. Mr Revell has roughly estimated this at £2,000,000,000, making a total of £6,500,000,000 for personal settled property. These estimates, it should be pointed out, relate only to persons domiciled in Britain. As we show in a later chapter, a further addition should be made for trusts established abroad.

It is obvious that the total of personal wealth in Britain is much larger than previous estimates have allowed for, and it is equally clear that many of the developments discussed in this chapter have been exerting a profound effect on its distribution. For one thing, there is the remarkable and unexplained fall in the 1950's in the proportion of settled property paying duty to total dutiable property—a significant fact thrown up by Mr Revell's studies. For another, there

---

[1] Revell J. R. S., 'An Analysis of Personal Holders of Wealth', British Association for the Advancement of Science, 1960. See also the *British Tax Review*, May-June 1961, p. 177.

[2] Lydall H. F. and Tipping D. G., *op. cit.*, pp. 92–3.

[3] Revell J. R. S., *op. cit.* (1961), p. 177.

[4] Revell J. R. S., *The Times*, July 11, 1961.

is the evidence he has produced which suggests that the chance of possessing an estate of a given size has increased more for younger people than for older people since 1926.[1]

These studies are raising fundamental questions about the extent of our knowledge concerning the redistribution of personal and family wealth in Britain in recent decades. They are equally relevant to the study of income distribution, particularly when account is taken of the unit of time, the unit of income and the concept of income. The questions that are asked become more rather than less important as a result of the new powers under the Variation of Trusts Act, 1958, and the Trustee Investments Act, 1961, allowing trustees to invest in equities. The number of applications to the courts for the varying of trusts has already made public some illuminating facts concerning the value of settlements.[2]

The potential uses of these trusts are, therefore, very great. They are often employed, according to Messrs Potter and Monroe, in the reorganization of the shareholdings in a family company with a view to reducing estate duty.[3] They are becoming increasingly popular as a means of splitting and spreading income and capital over the life of a family, born and unborn, for several generations.[4] In combination with other provisions in the tax planners' armoury, they may create income out of capital, transfer income from one person to one or more other persons, turn income into capital, and transform un-earned income into earned income. The substantial surtax advantages now given to earned income under the Finance Act, 1961, suggest various ways in which the latter may be effected. One method may be to pay large amounts of 'remuneration'—as 'earned' income to trustees; another to reduce profits by increasing director-owner remuneration.[5]

Discretionary trusts may also be designed to benefit a considerable number of persons—what the law describes as 'a wide class'. A particular instance of this type of covenant was referred to in the Board's Memorandum of Evidence to the Royal Commission. It provided for an annual payment of some £14,500 per annum on

[1] Revell J. R. S., *op. cit.* (1960).

[2] Three cases before the Chancery Division since March 1959 concerned trusts valued at over £250,000, over £1,500,000 and 'some' £2,300,000 (*The Times*, April 11, May 1, 1959, and January 13, 1961). See also Price L., 'The Variation of Trusts Act, 1958', which points out that the machinery of varying trusts under the Act 'is now of the greatest interest to all tax-planners' (*British Tax Review*, January-February 1960, p. 48).

[3] Potter D. C. and Monroe H. H., *op. cit.*, pp. 170 and 182–6.

[4] Potter D. C. and Monroe H. H., *op. cit.*; see, for example, Precedents No. 24 and 25 on pp. 185 and 195.

[5] See Bowyer J. L. M., *British Tax Review*, May-June 1961, pp. 149–50, and Wheatcroft G. S. A., *British Tax Review*, March-April 1961, p. 93.

G

discretionary trusts in favour of about fifty named beneficiaries, mainly children of the covenantor's relations or friends.[1] How wide the class may be in its social characteristics was illustrated in *Re Gestetner Settlement* (1953) Ch. 672. The settlor desired to benefit a class including four named persons, any issue living or thereafter to be born who was a descendant of two named persons, any spouse, widow or widower of any such person, five charitable bodies, any former employee of the settlor or the widow or widowers of such former employee, and any person for the time being the director or employee of certain named companies or the wife or husband or widow or widower of such person, or any director or employee of any company of which the said directors of the named companies were also directors. Obviously, this is a wide class, the membership of which could not be easily ascertained at any particular time. The settlor therefore defined the wide class, defined also a narrower class, namely, issue of the settlor, and defined a perpetuity period. He settled the capital of the trust fund upon trust for such of the wide class as the trustees should appoint, and in default of appointment upon trust for the narrow class. He then settled the income of the unappointed part of the trust fund, until the perpetuity day, upon discretionary trusts for the wide class, and subject to the exercise of such discretion upon trust for the narrow class. In short, he settled his property upon discretionary trusts and powers for a wide class, not all of whose members would be at all times ascertainable, with a gift in default to a narrower class all of whose members would at all times be ascertainable. It was held that the powers and trusts were valid.[2]

It would seem that, aided with the powerful weapon of the discretionary trust, there is no need to quarrel with the statement of one authority on the subject of tax planning: 'as far as estate duty is concerned, the Revenue benefit almost exclusively from the unlucky, the ungenerous and the unwise', and that anyone 'who is prepared to divest himself of most of his assets and does so with proper advice should not—given luck—trouble his executors with estate duty problems'.[3]

Edmund Burke might have repeated today what he said in his *Reflections on the Revolution in France*: 'the power of perpetuating our property in our families is one of the most valuable and interesting circumstances belonging to it and that which tends the most to the perpetuation of society itself . . . the possessors of family wealth

---

[1] Memorandum 119, *Vols. of Evidence, Royal Commission on the Taxation of Profits and Income 1952–5*.

[2] See also 'Certainly Not Uncertain', *British Tax Review*, March-April 1961, pp. 144–5.

[3] Grundy Milton, *op. cit.*, p. 122.

... are the natural securities for this transmission'. And we could equally well apply, with suitable modifications for the development of discretionary trusts today, Pollock's summing up on family settlements in 1894: 'there is nothing, perhaps, in the institutions of modern Europe which comes as near to an *imperium in imperio* as the discretionary settlement of a great English fortune. The settlor is a kind of absolute lawgiver for several generations; his will suspends for that time the operation of the common law of the land and substitutes for it an elaborate constitution of his own making.'[1] Yet, for this century at least, no statistical light has been cast by the Board of Inland Revenue on the effect of all these settlements, trusts and charities on the redistribution of income and wealth. When it comes to statistical illusions, there is no 'rule against perpetuities'.

[1] Pollock F., *The Land Laws*, 1894, p. 112.

CHAPTER 6
Statutory Income and Real Income

In the previous chapter we dealt with some of the chief methods by which income and capital are merged, transformed, divided and spread over time amongst members of the family and the wider kin group. A substantial proportion of these claims on current and future resources by a relatively small section of the population escapes taxation and estate duty wholly or partly and, in conventional terms of yearly income, is reflected either fractionally or not at all in the Board of Inland Revenue's income tables. The pattern of distribution as between income units in different ranges of income is thus changed; by omissions, by the creation of new units, and by rearrangements which depress the values of some units and raise the values of others.

We turn now to consider a complex group of factors which affect in similar and dissimilar ways the Board's tables. We deal first with those which mainly take the form of untaxed realized capital gains and capital receipts (not falling within the definition of taxable income); and, secondly, with a further and miscellaneous assortment of tax and estate duty avoidance factors. Chapter 7 considers the effects of pension schemes, and Chapter 8 deals with benefits in kind. Once again, the object is not to measure directly the effect of all these factors on the distribution of income and wealth, but to provide some indication of their actual and potential influence in changing the picture presented to us by the official distribution statistics.

The problems here considered are so diverse and the data so complex that, in this chapter, we shall attempt little more than a catalogue arranged under a number of broad headings. No claim is made to comprehensiveness; only the more important sources of non-taxable receipts and benefits are enumerated.

The nature of the subjects dealt with inevitably brings us back once more to the problem of defining income. This was briefly discussed in Chapter 2, and we saw that the definition used by the Board in classifying personal incomes was:

Income before tax is all the income brought under the review of the Department, after certain deductions. It is after deducting losses and capital allowances in the case of profits and professional earnings and the allowance for repairs in the case of income from property; it is also after deducting National Insurance and superannuation contributions and other allowable expenses, mortgage interest and similar annual payments. It is before deducting the personal allowances or life assurance relief. It excludes income not subject to tax, such as interest on National Savings Certificates, National Assistance grants and certain National Insurance benefits and grants (unemployment, maternity, sickness, industrial injury, etc.)[1]

For all non-census years the statistics for personal incomes by ranges are based on *income charged*. The information obtained from time to time through a census or sample survey is based on *income returned*. The precise statistical differences which result from the application of these different bases for determining the distribution of incomes have not been illuminated in the Board's reports. Nevertheless, it seems clear that the former may relate to income arising in different years—or over a period longer than a year—while the latter relates to income arising in a specified year.[2] Presumably the former includes, therefore, what are known as 'back duty assessments'. The sums involved in such cases may be substantial on account of income tax and surtax charged in a particular year being unsettled, under appeal, adjusted, discharged or remitted. Moreover, the amount of back duty work varies from year to year according to the staffing situation in the Board's offices and other causes. Some of this work is related to under-assessments resulting from fraud and evasion of taxes on income and profits. In 1939 there were 2,774 such cases involving £3,131,410 (total charge raised). In 1955 the number had risen to 19,663 involving £20,587,922.[3]

In what particular and precise ways all these factors have affected the Board's classification of incomes statistics since 1938 is not known. No information has been published, for instance, on the effects of back duty assessments. It is a fair assumption, however, that, if brought fully to account, they are more likely to have altered the statistical picture at the higher than the lower end of the income scale. It also seems unlikely that, whether adjustments are or are not made over a period of years, these factors would result in any over-

---

[1] *BIR 103*, p. 74.

[2] Two different years may be involved but they are both specified single years. For example: the 1954–5 sample survey covered wages, salaries, dividends, etc. arising in the year 1954–5, and profits and professional earnings arising in the trader's accounting year ended in 1953–4 (*BIR 99*, p. 66).

[3] *BIR 83*, *99 and 103* and *Report of the Comptroller and Auditor General*, Appropriation Accounts, 1959–60.

statement of incomes; the reverse is more probable. In general, therefore, we may say that the net result of these processes of treating income is more likely to err on the side of understating than over-stating degrees of inequality.

There we must leave this particular matter, however, and move on to consider certain official definitions of income. For convenience of description we shall henceforth refer to these definitions as 'statutory income'.

The main problems we now have to examine, and which constitute the subject matter of this chapter, may be phrased in three questions, namely:

(i) What is 'actual income' (as distinct from statutory income) and where does the line fall between the former and realized capital gains?

(ii) What have been the differences since 1938 between 'actual income' and statutory income, and what effect would such differences have on the income distribution statistics if all the appropriate adjustments were made?[1]

(iii) What have been the effects since 1938 of the deductions and exclusions from 'income brought under review' (referred to above in the definition quoted from the Board's report)?

In asking these questions we are, in a sense, trying to distinguish between the real world of income-wealth; the Board's concept of 'income brought under review' before making any deductions; and the effect of the deductions and exclusions from income reviewed.

In all three spheres it is necessary to remember that changes take place, and that these should be examined in any study which purports to investigate trends in the distribution of personal incomes over time. For example, what is reviewed and is not reviewed by the Board, and what is deducted and is not deducted, may change from year to year. Changes in the law, in extra-statutory concessions and in administrative practices may all be involved. This warning is necessary because, as we have shown in earlier chapters, the Board itself—as well as many students who interpret its statistics—draw conclusions from its tables which go far beyond a simple comparison of uniform data on 'income reviewed' or 'revenue work done'.

One further warning is needed before we begin to look at these three problem areas of defining income. This is relevant if we wish to distinguish separately the effects of wives' earnings on the distribu-

---

[1] The term 'statutory income' was used by Mr Shirras and Dr Rostas. In their study they made the assumption that 'actual income' and 'statutory income' were identical. While pointing out that this was not always the case, as a certain amount of income escaped taxation, they concluded that the error involved in the assumption was 'probably small' (Shirras G. F. and Rostas L., *The Burden of British Taxation*, 1942, p. 71).

tion of incomes by ranges. In doing so, account must be taken of the fact that in the Board's classification of wives' earnings, statutory income is differently defined. It is *before* the deduction of National Insurance and superannuation contributions and 'other allowable expenses'. 'It excludes Schedule E (employment) income from subsidiary sources if a taxpayer has Schedule E income from more than one source at the same time. It also excludes Schedule E income for which no tax deduction card is issued, such as family allowance and National Insurance pensions. The remuneration is the total Schedule E income from the principal source; when there is a change of job it is the sum of the earnings in successive jobs. It includes some salaries and fees where the main source of income is Schedule D.'[1]

When this definition is used in the classification of wives' earnings by type and range it is difficult to know what the results mean, particularly as there were about 1,500,000 'missing' earning wives (omitted or understated in terms of income in the husbands' tax returns).[2]

We now return to the definition cited earlier, which is used by the Board in its classification of total personal incomes for all income units. We may begin by examining, in particular, the effects of various deductions and exclusions before statutory income is calculated. In this and the following two chapters we single out some of the quantitatively more important items.

As it stands, the Board's definition may be thought by the layman to be vague and imprecise in certain respects. While National Assistance grants are specified, 'allowable expenses' and 'similar annual payments' are not. These, however, have a fairly precise meaning in the statutes (and to some extent also in administrative practice) and it would no doubt have entailed a much longer definition if the Board had attempted to specify them in detail. Nevertheless, we must explore their effects.

The Central Statistical Office, in its annual *National Income and Expenditure* tables, 'feeds' back into the Board's statistics of the distribution of personal incomes certain of these deductions and allowances for all income units including wives.[3] Its definition of personal income is thus, as we shall see, somewhat wider than the Board's operative definition. It includes as additions to income covered by the Board's data:

(i) Incomes below the exemption limit, except the incomes of persons receiving less than £50 a year.

(ii) National Assistance grants and those National Insurance benefits and grants which are not liable to tax (unemployment,

---

[1] *BIR 100*, p. 80 and *BIR 101*, p. 70.
[2] See Chapter 4.
[3] See *NIBB*, 1960. Notes on table 22, p. 68.

sickness, maternity, injury, disablement and death benefits and grants).

(iii) Income in kind of domestic servants and agricultural workers.

In its book on 'Sources and Methods' the Central Statistical Office explains that because no information is available about the distribution of the beneficiaries by income ranges it is not possible to feed back other categories of personal income which are directly received by (or imputed to) individuals.[1] However, the additions they do make for (ii) and (iii), though explicable on other grounds, have the effect of introducing further inconsistencies into the personal income distribution data. For example: while compensation for loss of office, which applies particularly to the higher income groups, is not included owing to the lack of information, unemployment benefit is included. The same point may be made about sickness benefit. While personal benefits provided by employers for their higher paid employees (medical care, private hospital and nursing home treatment, private sickness benefit, convalescent holidays, travelling and expenses on sick leave) will generally not be counted or counted in full, sickness benefit under National Insurance is included. Income in kind received by two low income groups (agricultural workers and domestic servants) falls within the Central Statistical Office definition of personal income; the infinite variety and greater absolute value of income in kind received by many higher paid employees is excluded insofar as it is not fully assessed to tax at current market prices. This, as we shall see later, is unlikely to be the case.

Estimates of these two items (ii and iii) have, according to the Central Statistical Office, 'been allocated to what seem to be the most appropriate income ranges'.[2] Without an immense amount of work it is difficult to calculate the effects of these allocations, described by Mr Lydall as 'comparatively straightforward adjustments'.[3] However, it is clear from inspection of the comparative data that the personal incomes before tax of the lower paid employees have been adjusted upwards whereas the income of those in the higher ranges remains unchanged. In other words, the effect of applying the Central Statistical Office definition of income is to make the statistics of income distribution look more equalitarian than those produced by the Board's definition. This is not to say that these sources of income, which chiefly affect the bottom and middle groups, should not be included; it is the exclusion of other categories of income, chiefly attributable to the top groups, which produces a spurious equalitarian effect.

When, however, we examine the adjustments (or feed back) made

[1] *National Income Statistics: Sources and Methods*, 1956, p. 69.
[2] *NIBB*, 1960, p. 68.
[3] Lydall H. F., *op. cit.* (1959), p. 4.

by the Central Statistical Office to the 1937–8 income distribution data we find that they produced a quite different effect from that which followed adjustments in the 1950's. Similar additions were made to the earlier data in respect of categories (i), (ii) and (iii) above.[1] The results are set out in Appendix C which includes for purposes of comparison the corresponding data from the Board of Inland Revenue's table for 1937–8.

Some of the differences that emerge are, on the face of it, surprising. They make the pre-war distribution look more unequal. Thus, while the figures are, in round totals, the same for the range £2,000–£3,000 and £5,000 up, the Blue Book records about 1,650 more 'incomes' in the range £3,000–£5,000 with a total of £7,000,000 more income (average of £4,242). We can only conclude that this 'adjustment' covers an unpublished discrepancy in the Board's statistics. Moving down from the income range £1,500–£2,000 it is noticeable that the number of adjustments increase, both in respect of the number of 'incomes' and the aggregate amount of income involved. Considering the very restricted scope of public assistance, unemployment and other pre-war benefits, to say nothing of income in kind for domestic servants and agricultural workers, it is remarkable that such a large number of 'incomes' between £500 and £2,000 have been affected by this adjustment. The Appendix shows that, in total, the Blue Book records 160,000 more 'incomes' and £118,000,000 more income. Of the latter figure, £89,000,000 is attributed to the income ranges above £500. Over one-quarter (or £32,000,000) is attributed to incomes above £1,000. It is impossible to believe that this is caused by additions in respect of public assistance grants and income in kind of domestic servants and agricultural workers.

Some investigators, we may note, have worked from the Board's figures; some from the Blue Book. These adjustments or 'discrepancies', which no private investigator has commented on, do not increase confidence in the reliability of the critically important data for 1938. The Board has never published any corrected figures or referred to these substantial adjustments. The reader has been left to assume that the original data still stand.

Finally, to complete this examination of the two sets of official data relating to income distribution we may note that the following types of income which, as the Central Statistical Office states, all form part of personal income for the national global figures, are excluded from its definition for the classification of income by ranges:[2]

---

[1] *NIBB*, 1958, p. 74.
[2] *NIBB*, 1960, pp. 68–9. See also *National Income Statistics: Sources and Methods*, 1956, p. 69.

(a) Income in kind other than that of domestic servants and agricultural workers.

(b) Certain non-taxable grants from public authorities (milk and welfare foods, grants to Universities and schools, etc., and industrial services for the disabled).

(c) Investment incomes of non-profit-making bodies and of life assurance and superannuation funds.

(d) The amount by which the imputed rent of owner-occupied dwellings exceeds the Schedule A valuations.

(e) Any other differences between the incomes shown in the Inland Revenue returns and the corresponding estimates included in personal income.

(f) That part of the income of self-employed persons which is covered by allowances for depreciation.

(g) Accrued interest on National Savings Certificates.

(h) Post-war credits.

(i) Incomes of persons receiving less than £50 a year.

(j) Employers' and most of employees' contributions to national insurance and superannuation schemes.

According to the government statisticians, this list represents the main items of personal income not allocated to particular ranges of income. In total, it was estimated to amount to £615,000,000 out of an aggregate national total of £5,078,000,000 for 1938 or 12 per cent. In 1959 the respective figures were £3,261,000,000 and £19,676,000,000.[1] On the assumption, therefore, that the estimates of total personal income are broadly correct we can say that the unallocated portion has steadily risen since 1938 and in 1959 stood at 17 per cent. Included in this portion (and in the foregoing list) are several items which it may not be proper to allocate to personal income by ranges; for example, employers' contributions for national insurance. There are other major items, like capital gains and the proceeds of insurance policies, which could be considered as income that are excluded from the Blue Book income figures. Much depends, of course, on the definition of income adopted in considering whether or not personal income should include the benefits of social service expenditure in kind, undistributed company profits, non-taxable capital gains, and many of the categories of benefits and increments of spending power analysed in this study. These questions all form part of the more fundamental problems briefly discussed in Chapter 2 of the curtain between income and capital.

In the rest of this chapter we return to consider in more detail certain elements of these problems. In particular, we shall be concerned with the growth and present order of magnitude of some of the more important sources of spending power which do not enter into

---

[1] *NIBB*, 1958 and 1960, table 22.

the concept of statutory income. By the time we have finished it will be obvious why we have made no attempt to quantify these factors as a whole. So little is known and so formidable are the difficulties that the government statisticians have not attempted to do so for the types of income listed above.

In his study, Mr Lydall discussed some of these problems relating to unallocated income, and hazarded various estimates which he then applied to the basic data for the distribution of incomes by ranges. We shall refer to these later as we examine some of the main components of non-statutory income and wealth.

Capital Appreciation and Realized Gains

The question of capital gains was considered by the Royal Commission in its *Final Report*. In the Report and in the accompanying Memorandum of Dissent by Messrs Woodcock, Bullock and Kaldor, the problems of definition and of distinguishing between capital appreciation and realized gains were discussed at length. While it is unnecessary to explore these matters in detail here it may, however, be pointed out that any statistics which essay to portray changes in the distribution of personal incomes should take account of realized capital gains of a regular nature; in other words, untaxed additions to spendable income. This needs to be done, irrespective of opinion as to whether the additions should or should not be taxed, if we wish to obtain a more accurate picture of the distribution by ranges for a given period of time. The principle is essentially the same as that adopted by the Central Statistical Office in making adjustments for the income in kind of domestic servants and agricultural workers.

The evidence presented in this study and in the Report of the Royal Commission makes clear that since 1938 the factor of realized capital gains has greatly increased in importance. It affects, in particular, the incomes attributed to the top 2 per cent of incomes in the Board's statistics. As Mr Brittain has shown, the changes in the percentage shares of various income groups between 1938 and 1955 were 'due entirely to the behaviour of very high incomes'.[1]

Writing in 1954, the authors of the Memorandum of Dissent gave some indication of the magnitude of the problem of capital gains at that time. Discussing the relationship of capital gains to tax avoidance they said:

The full significance of the omission of capital profits from taxation only becomes clear, however, when it is appreciated that the extent to which rewards take the form of tax-free capital

[1] Brittain J. A., 'Some Neglected Features of Britain's Income Leveling', *American Economic Review*, Vol. L, No. 2, May 1960, p. 596. This study is described in more detail later in this Chapter.

gains rather than taxed dividend income is not something that is fixed by Nature, but is very much subject to manipulation by the taxpayer. A community with a highly developed capital market like Britain offers wide opportunities for an individual so to arrange his affairs that the accrual of benefits from his ownership of capital takes the form of capital appreciation instead of taxable income. It is well known (indeed it is constantly broadcast by the financial press) that it 'pays' a surtax-payer to select securities which have a low dividend yield but a high degree of expected capital appreciation, owing to rising dividend payments over time; to purchase bonds that stand at a discount, or bonds that can be bought in an 'unassented' form (like German and Japanese issues at the present time) so that the value of the interest that is periodically paid on them appears as an appreciation of capital, and not as taxable income; or to convert income into capital appreciation in innumerable other ways.

The sale of securities 'cum dividend' and their re-purchase 'ex dividend' can normally be accomplished through the Stock Exchange on terms that relieve the holder of surtax if not the whole of income tax. Indeed through transactions between parties one of whom is taxed on capital gains (as is the case with stock-jobbers or finance houses) whereas the other is not, or one of whom is exempted from tax altogether (as with pension funds) while the other is not, it is always possible for taxable income to be converted into capital gains at the cost of the revenue. For the stock-jobber can offset his dividend or interest income by a capital loss; the pension fund can reclaim tax on interest and dividends received. It pays therefore both stock-jobbers and pension funds to buy securities cum dividend and to sell them ex dividend so long as the capital loss on the whole transaction is less than the gross dividend or interest payment; it pays the ordinary investor to sell cum dividend and to buy ex dividend so long as his net capital gain is greater than the net dividend or interest (after deduction of income tax and surtax) which he sacrificed. The existing legal provisions against 'bond-washing' prohibit this practice when the sale and repurchase is accomplished 'by the same or any collateral agreement'. But there is nothing illegal in buying and selling securities through two separate and unrelated transactions, and it would indeed be impossible to frame anti-avoidance provisions which prevented a man from selling one day and buying the next day, especially since he need not even buy back the identical security to accomplish his object.

An extreme example of the conversion of taxable income into tax free capital gains which has come into vogue in recent years consists of the device known as 'dividend stripping'. This concerns companies (mainly private companies) which have accumulated large balances of profits; such profits having been subjected to income tax are available for distribution as a taxed dividend. If such companies are bought by a finance house at the time when the companies' assets have been converted into cash or financial investments it is possible to extract the accumulated reserve by means of a special dividend and to sell the residual assets immediately afterwards at a corresponding loss. Since the finance house's receipt of the (net) dividend income will be offset (and may be more than offset) by its loss on the capital transaction the result is that an amount corresponding to the current income tax on the capital loss will be reclaimable from the Exchequer. Depending on the terms on which the original owner sold the company to the finance house, the gain from the repayment of income tax is shared between the finance house and the original owner, and amounts to a greater or less proportion of the income tax deemed to have been deducted at source from the special dividend. An analogous result is attained (with even greater effectiveness) if the company is sold to a charity or a pension fund which can reclaim the whole of the tax on the dividend by virtue of its tax exemption; and which makes it profitable to pay a price well in excess of the realizable assets of the company. No doubt extreme practices of this kind could be stopped by special anti-avoidance provisions. But in our view so long as capital gains remain exempt from taxation it is impossible to deal with the problem of the conversion of income into capital gains in all its possible forms by specific pieces of legislation.[1]

All these opportunities, moreover, are far more readily available to the large property owner than to the small saver; and it is well known from American experience that capital gains are a major source of large incomes but unimportant as a source of smaller incomes. The Report, while it pays a great deal of attention to the alleged inequities that the taxation of capital gains would involve as between one property owner and another . . . makes no mention of the far more serious inequities that arise as a result of the exclusion of capital gains from taxation as between those who can (and do) manipulate their affairs so as to augment their tax-free gains at the cost of their taxable income and those who are not in a position to minimize their tax liability in this manner. In fact the Majority find no occasion to refer to problems of tax avoidance in connection with capital gains at all.

With regard to the capital profits associated with ordinary shares, the Majority admit that as companies are growing through the steady retention of part of their current profits and the consequent growth of their earnings and dividends, 'a process is detected which has the effect of adding to the capital of a shareholder through savings made out of the company's income without his share of the savings ever having been subjected to surtax as income in his hands'. They argue, however, that since undistributed profits bear the whole of the profits tax as well as income tax at the standard rate, even though this charge 'has no ascertainable relation to

---

[1] Since this was written, legislative action has been taken against some of the devices described in the Memorandum of Dissent. See also note on p. 143.

the benefit that some shareholders may obtain from the fact that their share of retained profits is, in effect, saved for them in the corporate pool', 'it is impossible to ignore the countervailing circumstance that profits made in corporate form do in fact bear a supplementary charge which is not imposed on other forms of profit or income'.

This argument implies that undistributed profits are the source and the measure of the appreciation of ordinary shares so that the case for taxing the appreciation of these shares arises from the difference between the tax actually paid by the companies on undistributed profits and the tax that would have been payable if the aliquot share of undistributed profits had been imputed to the individual shareholders. This approach completely overlooks the fact that in the case of successful companies the growth in market capitalization resulting from the growth in earnings and dividends may greatly exceed the growth in the companies' reserves through the continued ploughing back of profits. To show how important that fact can be it is sufficient to refer to a few leading equities. In the case of Woolworth, for example, the market capitalization of the ordinary capital amounted (in February 1955) to some £m280 whereas the share capital and accumulated reserves together amounted to £m39. In the case of Marks and Spencer the market capitalization amounted to £m128 whereas capital and reserves amounted to £m20. In both these cases therefore 83 to 86 per cent of the accrued appreciation in the value of the shareholders' capital has never borne any form of tax, however indirect; nor can that appreciation be described as bearing the character of isolated, unexpected or non-recurrent gains, considering that it has accrued over a long series of years—in fits and starts no doubt, with fluctuations in the sentiment of the capital market, but unmistakably as a by-product of the growth in the companies' profits and dividends.[1]

Taking a longer view over the whole period of the development of progressive taxation the gain in the value of equity investments has been striking, even though no allowance is made in the following example for price changes. A survey undertaken by a firm of London stockbrokers in 1959 showed, for example, that the total value of holdings of six Ordinary shares valued at £1,000 each in 1913 had risen by September 1, 1959, to £73,840.[2] It was also pointed out that

[1] *Final Report*, Cmd. 9474, 1955, pp. 368–71. See also discussion of accrued income and realized income in Mr Kaldor's *An Expenditure Tax*, 1955, pp. 37–41.

[2] During the same period the internal purchasing power of the £ fell to about one-quarter. The six companies were J. & P. Coats, Distillers Company, Guest Keen and Nettlefolds, Vickers, Peninsular and Oriental Steam Navigation Company and Courtaulds (also reported in *The Times*, September 15, 1959). If £1,000,000 had been invested on January 1 each year since 1919 in a representative group of 'blue chip' industrial equity shares and the gross income

if £1,000 had been invested in 16 different companies and no fewer than 10 of them had gone bankrupt with worthless shares the fund would still have £16,000 (the original investment) in market value at 1920 if the remaining six companies were those shown. Taken still further to September 1959, the value of £73,840 would have permitted no fewer than 67 completely worthless investments out of 73 made in 1913.

What effect would realized capital gains resulting from capital appreciation of this order have had on the income distribution statistics if such gains had been treated statistically as income? More pointedly, what has been the total of annual realized gains since 1937–8, and how should such gains be distributed amongst the different income groups?

Without an immense amount of research it is impossible to hazard an answer to such questions. In any event, the statistical problems involved would be formidable. Irregular and occasional gains might have to be treated differently from regular and more secure realized gains.

In 1951 the Board of Inland Revenue estimated (for the purposes of gauging the long-term yield of a capital gains tax) that annual average capital appreciation would amount to £150,000,000. It was assumed by the Board that in the long run net realized gains correspond to net appreciation.[1] Three years later, in the light of changes in policies and the economic situation, the Board revised its estimates to £200–250,000,000.[2] Mr Kaldor and his co-authors, in their Memorandum of Dissent, argued, however, that these estimates were far too low. After analysing the long-term trend of profits and capital values they said:

We believe that the rate of growth of industrial production in this country cannot be put at less than 3 per cent a year. On this basis the long-term rate of growth of profits is unlikely to be below 3–5 per cent a year. The lower of these estimates assumes that the inflationary trend in prices experienced since the war will be completely eliminated in the future. The upper estimate also assumes that Governments will succeed in controlling inflationary trends more successfully in the future than in the past; but it takes into account the possibility of the continuance of a moderate degree of price inflation. Assuming that dividends keep pace with the rise in earnings, this implies an annual increase in dividend

reinvested at the end of every year the total value of the fund on January 1, 1950, would have been £122,080,000 and on January 1, 1960, £646,330,000. Allowing for capital appreciation or depreciation as well as gross income, the overall yield would have been 10·6 per cent per annum (*The Times*, March 24, 1960).

[1] *Final Report*, Cmd. 9474, 1955, p. 378.
[2] *Ibid.*, p. 378.

payments of the order of £m25–£m40, and an average increase in the value of ordinary shares of £m500–£m800 a year.

The percentage rise in the value of real property cannot be estimated with any such confidence but is not likely to be as great. Assuming an increase of the value of real property of the order of 1–2 per cent a year and taking into account capital gains on the sale of unincorporated businesses, capital appreciation in forms other than ordinary shares cannot be put at less than £m100–£m200. Hence the long-term rate of capital appreciation in this country in all forms should be put at a minimum of £m600–£m1,000 a year.[1]

This was written in 1954 before the great boom of the 1950's in property and land values and when share prices were still a long way below the levels reached in 1960. Between March 1958 and March 1960, for instance, the value of securities quoted on the Stock Exchange rose by no less than £12,000,000,000, or over 36 per cent.[2] By contrast, the rise during the four years ending March 1954 (which was taken into account by Mr Kaldor and his co-authors) was only £5,000,000,000.[3] Not until after 1953 did equity prices begin to rise steeply. By 1960, the gross dividends on the shares included in the *Financial Times* Index were more than three times as high, and the capital value nearly three times as high, as in 1945.[4]

During the 1950's property shares as a whole appreciated at an even faster rate. According to *The Times*, quoting a leading firm of stockbrokers in 1960, the rise was of the order of 900 per cent in nine years.[5] New office building was then at the rate of over £70,000,000 a year. *The Economist* reported that to make a capital gain of less than £200,000 on a £1,000,000 speculation was considered, in 1959, to be unsatisfactory.[6] 'Never,' wrote Mr Frederick Ellis, City Editor of the *Daily Express*, in 1959, 'has so much money been made from property as in the decade 1949–59.'[7] He estimated that nine men had made a total net capital gain of £40,000,000 since the war. The Board of Inland Revenue's statistics, however, show that in 1957–8 there were only 2,600 'income units' with incomes before tax exceeding £20,000 a year; the average income was only £32,308.[8]

Mr Douglas Jay, in surveying the great increase in capital gains during the 1950's, drew attention to the important effects of full

[1] *Final Report*, Cmd. 9474, 1955, pp. 380–1.
[2] *The Times*, May 8, 1960.
[3] *Final Report*, Cmd. 9474, 1955, p. 380.
[4] *Investors' Chronicle*, December 23, 1960, p. 1173.
[5] *The Times*, August 29, 1960.
[6] *The Economist*, May 16, 1959, p. 606.
[7] Quoted in *New Statesman*, June 27, 1959.
[8] *BIR 102*, p. 72.

employment.[1] He concluded that an 'almost wholly new force, making powerfully for greater inequality, is also almost certainly now starting to operate, due ironically to the full-employment economy itself'. This, he said, was 'the growingly important new phenomenon of steady long-term capital gains and rising dividend incomes in the hands of the minority who hold equity shares'.

Combined with various categories of realized tax-free capital gains, the non-distribution of corporate profits explains in some measure the remarkably low yields of surtax during the 1950's. Whether and to what extent undistributed profits do form part of personal income is a much debated issue.[2] There seems to be no doubt, however, that they do insofar as they lead to realized capital gains.

Undistributed profits showed a substantial rise during the Second World War. If all pre-tax undistributed income is imputed to individual shareholders, the 1938 and 1949 personal income distributions show no levelling effect.[3] Between 1949 and 1959 the yearly total of undistributed income[4] of all non-nationalized companies rose from £914,000,000 to £2,147,000,000 or 135 per cent. Yet, during the same period, the annual investment income of surtax payers increased by only £235,000,000 to £610,000,000 or 63 per cent despite the lowering of the surtax exemption limit due to inflation.[5]

Although the difficulties of apportioning investment income between different classes of taxpayers are very great Mr Lydall, in his paper on incomes, attempted an estimate of the share of the top 1 per cent of income units in shares or dividends (and hence in undistributed profits). He put the figure at 77 per cent in 1938, 57 per cent in 1949 and 49 per cent in 1954.[6] In a later study with Mr Tipping on the distribution of personal wealth, however, he found that in 1954 the top 1 per cent owned 81 per cent of stocks and shares in companies.[7] While the methods employed in arriving at these two sets of data are necessarily somewhat different, nevertheless, the discrepancy is a large one. Without much more basic information about both unearned income and wealth this discrepancy is unlikely, however, to be satisfactorily resolved.

[1] Jay Douglas, *Socialism in the New Society*, 1962, especially pp. 30–1 and 178–215.

[2] See discussion on Mr Lydall's paper, *op. cit.* (1959), pp. 37–46.

[3] Brittain J. A., *op. cit.*, p. 597. This study is discussed in more detail at the end of this chapter.

[4] After taxation but before providing for depreciation and stock appreciation (*NIBB*, 1960, table 26).

[5] *BIR 103*, table 72.

[6] Lydall H. F., *op. cit.* (1959), Appendix p. 36. A more recent estimate suggests that all equity shares are owned by 5 per cent of the population. (Cummings G., *Stock Exchange Journal*, Summer 1960, p. 11.)

[7] Lydall H. F. and Tipping D. G., *op. cit.*, p. 90.

H

In his earlier paper, Mr Lydall argued that the proportionate fall since 1938 'may be explained by the growth in the importance of life and superannuation funds as investors in equities, by the decline in the position of the rentier compared with those with an earned source of income, and by the sub-division of estates in anticipation of death duties'. Even supposing that there was such a fall between 1938 and 1954 a further possibility would have to be investigated. So far as the top 1 per cent are concerned (including their families and kinship members) it may be that income-wealth is accruing to them in different forms and through different channels—life assurance, superannuation, tax-free lump sums on retirement, education covenants, discretionary trusts, income in kind and so forth. The benefits of equity investments, particularly in those companies whose shares have appreciated in a striking manner, can now accrue to a family (as distinct from an individual) in a variety of extremely complicated ways. It is much too simple to consider, as some writers do, only income in the conventional form of payments to individuals of interest and dividends.

The author of a taxation survey in *The Economist* in 1957 attempted to estimate the surtax loss resulting from the non-distribution of corporate profits.[1] Taking the gross undistributed profits of £3,000,000,000 in 1955 he assumed, first, that one-half of this was paid out in dividends and, secondly, that one-half of these dividends went to surtax payers. On this basis the total increase in surtaxable income would be £750,000,000 for 1955. If surtax were paid (as the author further assumed) at the rate of 3s 6d in the £—the rate then applicable to incomes between £3,000 and £4,000—the yield would be £131,000,000 or as much as the total surtax revenue actually raised in 1953–4.

More significant still, the addition of £750,000,000 to the total of personal income officially attributed to all those with incomes of £2,000 and over before tax in 1955 would represent an increase of approximately 55 per cent (from £1,360,000,000 to £2,110,000,000).[2] Even this would understate the rise because not all incomes in the range £2,000 and over would have been surtax payers. Moreover, the calculation assumes that only one-quarter of gross undistributed profits went to surtax payers in the form of dividends.

But whatever the effect on the share of income accruing to surtax payers or the top 1 per cent of incomes (a smaller group) it has to be remembered that statistical exercises of this kind relate only to the question of undistributed company profits and the lag in personal investment income. Such exercises leave out of account other sources and, as we describe them later, a variety of forms of untaxed capital

[1] *The Economist*, February 9, 1957, p. 489
[2] *NIBB*, 1958, table 31.

gains. In addition, it should be pointed out that it is a matter of extreme difficulty, as Mr Lydall and others have found, to identify the recipients of the investment income that is distributed. Some of these difficulties arise from the failure of taxpayers to report dividends and interest taxed at the source.[1] In the last annual report of the Board of Inland Revenue these omissions were estimated to amount to £260,000,000 for 1958-9.[2] In 1951-2 the estimate was £200,000,000.[3] The Board's annual tables classifying personal incomes by ranges therefore exclude such sums except insofar as partial adjustments are made following a sample survey of incomes as in 1949 and 1954. Even on these surveys 'considerable deficiencies' were noted.[4] The precise basis of the adjustments that the Board does make at intervals to take account of some unreported income is not, however, explained. Mr Brittain has pointed out that, between 1949 and 1955, the adjusted investment income total shown by the Board rose by only about 6 per cent whereas the national income estimates by the Central Statistical Office showed for the same period a 38 per cent gain in personal investment income.[5]

As we have already emphasized, the purpose of this limited excursion into the complex fields of capital gains, undistributed corporate profits, and unreported investment income is not to estimate precisely the effects on changes in the distribution of personal incomes since 1938. Without much more basic data, the task is an impossible one. This is made clear at the end of this chapter when we consider in more detail the adjustments made by Mr Lydall and Mr Brittain. Here we can only draw attention to the gaps in knowledge and to some of the major factors which should be brought to account in any comprehensive and detailed study of changes in the distribution of income and wealth.

Other Forms of Realized Capital Gains

Certain forms of capital gains and associated avoidance devices, though relatively small in terms of revenue forgone, are relatively large as benefits in the hands of individual income units and families. Thus, they may influence strongly the statistical behaviour of the top 1 per cent of income units.

For over twenty years attempts have been made by different

---

[1] Others may be found in the growth of investment income accruing to accumulation settlements—a point overlooked by income statisticians. Such income does not count as income of either the settlor or beneficiary and presumably is not returned by either. The Board, therefore, has no information and cannot 'adjust' its tables.

[2] *BIR 103*, p. 45.

[3] *BIR 96*, p. 49.

[4] *BIR 99*, p. 87.

[5] Brittain J. A., *op. cit.*, p. 596.

Governments to prevent or limit large individual gains resulting from a variety of activities—bond washing, dividend stripping, the use of one-man companies, the transfer of income abroad, the use of covenants, the sale of 'know-how', hobby farming, take-over retirement lump sums, and other devices. Most of these methods of legal avoidance are not just the product of the post-war years of high taxation. The debates on the Finance Bills of 1936, 1937 and 1938 provide a mass of evidence on the extent and nature of avoidance methods in use before the Second World War.[1] Appendix E to this study also furnishes some material illustrating the role of insurance in the growth of certain of these schemes.

Under a series of headings we now bring together other data which, assessed as a whole rather than as isolated facts, have a bearing on the statistics of income and wealth since 1938.

Dividend Stripping, Bond Washing and Other Transactions in Stocks, Shares and Securities

One writer in the *British Tax Review* estimated in 1959 that the revenue loss attributable to various forms of dividend stripping was between £4,000,000 and £10,000,000 over the period November 1955 to April 1958.[2] The Chancellor of the Exchequer, when discussing the effects of dividend stripping and bond washing in May 1960, said that the tax at stake 'was large, and was increasing'.[3] These devices, including some known as the 'strip trick', 'the scissors', 'stock-shunting', 'lay jobbing' and the 'forward strip', were described and illustrated in the *British Tax Review* in the same year.[4]

So far as bond washing operations were concerned, the Chancellor added that the Treasury was dealing with devices where the gains of the operators on individual schemes could run into many hundreds of thousands of pounds and the possible loss to the Exchequer ran into 'many tens of millions of pounds'. From the context it appears that this refers to an estimated annual loss. It is a reasonable assumption that 'many' means at least £50,000,000. The Chancellor complained that 'it is largely ineffective to legislate against new devices after they have been brought into operation. One series of operations is enough to bring huge gains, and when the legislation comes along it does not catch that particular series of operations'. One device called current dividend stripping was on the increase, and could lead to an enormous loss of tax revenue.[5] In the Finance Bill, 1960, the Government took

[1] See also *Report of the Royal Commission on the Income Tax*, 1920, Cmd. 615, especially pp. 135–9.

[2] Fletcher E., *British Tax Review*, November–December 1959, p. 424.

[3] *Hansard*, H. of C., May 3, 1960, Vol. 622, Col. 896.

[4] Potter D. C., *British Tax Review*, July-August 1960, p. 248.

[5] *Hansard*, H. of C., May 3, 1960, Vol. 622, Col. 894–5. Three years earlier the Stock Exchange Council in a circulated memorandum had specifically asked

new powers in yet another attempt to prevent transactions of this kind. These were subsequently modified in committee. This legislation was not retrospective, despite the warnings given for over twenty years.

Earlier attempts to deal with avoidance devices concerning stocks and shares had, according to *The Economist* in 1960, 'failed abysmally';[1] yet five years earlier the Government had optimistically considered that several clauses in its Autumn Finance Bill would be 'absolutely effective in our object of killing dividend stripping'.[2] Legislation aimed at bond-washing goes back at least as far as 1937.

The One-Man Company

The use of the separate legal personality of a corporation by individuals, families and professional people as a means of tax and estate duty avoidance has a long history, and has inspired a great deal of literature on tax planning. In 1939, for instance, Sir John Simon in introducing his Budget referred to the growth of certain one-man company schemes as a method of transforming taxable income into non-taxable capital. 'These schemes,' he said, 'of tax avoidance are so flagrant and are so deliberately devised to get round the legislation of 1936 and 1937 that I shall have no hesitation in recommending retrospective action . . . in accordance with the very clear warning I gave last year.'[3] Nevertheless, as the Report of the Royal Commission on Taxation in the 1950's showed, they continued to present serious problems to the Inland Revenue.

The essence of the matter is simple: incorporation (which, as *The Economist* has said, 'is too easy and too cheap')[4] can be used either to distribute taxable income over time and over the kinship group so as to minimize tax liability, or it can be used to transform the taxable income of the company into non-taxable capital in the hands of the shareholders.[5] Much can happen to it subsequently, as previous chapters have shown, to ease its disappearance from estate duty statistics. If only a relatively small proportion of top income

members not to undertake 'bond washing' business. It was said that the Council were preparing amendments to the Stock Exchange rules in order to prevent this sort of dealing (*The Times*, August 14, 1957).

[1] *The Economist*, April 30, 1960, p. 445.

[2] *Hansard*, H. of C., November 8, 1955, Vol. 545, Col. 1666.

[3] *Hansard*, H. of C., April 25, 1939, Vol. 346, Col. 993.

[4] *The Economist*, September 24, 1960, p. 1211.

[5] The brief summary of the methods of using incorporation for tax avoidance purposes given in the following pages may be supplemented by reference to the *Final Report of the Royal Commission* and the Memorandum of Dissent, Cmd. 9474, especially pp. 180–5, 306–7, 312, 314 and 388; Wheatcroft G. S. A., *The Taxation of Gifts and Settlements*, 1958; Grundy M., *Tax Problems of the Family Company*, 1958; and Stanford D. R., *Tax Planning and the Family Company*, 1957.

units converted themselves into one-man companies between 1937–8 and the 1950's the effect on the shape of the income distribution curve could be quite important.

The expression 'one-man company' is a legal contradiction, since a company must have at least two members. The term is used colloquially to describe a company which is effectively under the control of one person although both ownership and control may ostensibly be divided among a number of individuals, whether as shareholders or directors.

The main opportunities for tax avoidance offered by the formation of a one-man company may be illustrated by the hypothetical example of a business man X who formed a company in 1959 to take over his business which in that year showed a profit of £10,000. If he had not formed the company, the whole £10,000 would be taxed as his personal income for the year. Now, however, it would emerge in the first instance as income of the company, which could vote him a salary of £2,000 out of the £10,000 without rendering him liable to surtax. If smaller profits were expected in the future, the remaining £8,000 could be carried forward and voted to X in subsequent years. This would not prevent him from drawing the whole £8,000 out of the company at any time as an advance against future salary. Alternatively the £8,000 could be divided among X's children or other relatives so that none of them received enough to attract surtax. If the company showed a profit of more than £2,000 after deducting directors' salaries up to a certain limit, it would be liable to profits tax. To get over this, X might decide to form not one company but several.

These various devices have not been allowed to continue completely unchecked. The Revenue have certain powers to nullify their effect. These powers, however, are only invoked where tax avoidance was clearly the main object of the transactions in question and they could not reasonably be justified on other grounds. This is a very loose net through which large sums of tax escape.

The methods of splitting annual income described above may reduce considerably the total tax liability, but so long as the £10,000 profit is eventually distributed to individual taxpayers in the form of income it will still attract income tax and possibly surtax. There are, however, various ways in which it can be transformed, by the time it reaches X, into capital and thus removed from the category of personal income. The various ways of doing this cannot be described here in detail, but one device deserves special mention, since it was used on a large scale in the property boom of the 1950's. A building speculator would form (or take over) a group of private 'investment' companies to hold each of the properties in his possession. When he 'sold' a property the legal ownership would remain in the hands of the company. Only the shares would change hands and the profit on the transaction would be tax-free. In some cases a string of companies was formed to own a single block of flats. This particular weapon was thought to have been knocked out of the tax avoidance armoury by the Finance Act, 1960, after it had received wide publicity in connection with the affairs of H. Jasper & Co Ltd.[1]

Of the innumerable other methods by which one-man companies can be used to remove both income and capital from the reach of the tax collector, we can only list a few: by a company 'capitalizing' its profits and distributing them among the beneficiaries of a settlement in the form of non-taxable bonus shares or debentures;[2] by payments made by a company to an insurance company as premiums on a policy to which an infant was only contingently entitled and thus not taxable as income of the infant; by 'capitalizing' income because the Special Commissioners cannot apportion income by surtax directions to beneficiaries under a discretionary trust;[3] by the 'employment' for remuneration in these companies by directors of the wives of other directors;[4] by a gift of shares followed by a bonus issue as only the gifted shares attract duty and the bonus shares are free;[5] by the formation and subsequent liquidation of one-man companies by

---

[1] See the reports by Mr Neville Faulks and Mr R. K. Lockhead on H. Jasper & Co Ltd and the Pilot Assurance Company (HMSO, 1961). These reports, in investigating the use made of 161 companies by three men, concluded that breaches of the Companies Act 'are not only committed by companies and their directors but also connived at by the Banks and financial institutions' (p. 17). The use of subsidiary or investment companies in the field of property development was illustrated in 1959 by the case of one property owner holding 451 directorships (*The Times*, September 26, 1959). A specialist in the formation of companies stated at the City of London Magistrates' Court in March 1960 (*a*) that a property director had wanted to buy 100 companies, and (*b*) that his firm had supplied between 12,000 and 20,000 companies all over the world (*The Times*, March 22, 1960). See also report in *The Guardian* (January 2, 1962) of a company in London which has a stock of 'ready-made instant companies, all with impersonal names and a nominal authorized capital of £100, designed for the growing body of people who have realized the advantages of turning either themselves or their businesses into limited companies'.

[2] On the use of settlements in relation to property companies see Crump S. T., *British Tax Review*, March-April 1961, pp. 95–108.

[3] See Stanford D. R., *op. cit.*, especially Precedent No. 17, p. 230.

[4] The Finance Act, 1959, increased substantially the maximum amounts which may be deducted for directors' remuneration by director-controlled companies for profits tax purposes (Section 33). Directors who appoint members of their own families as directors and provide for the fees by reducing their own remuneration or dividends gain tax advantages; surtax is saved and the family receives personal and earned income reliefs. See also *Minutes of Evidence, Royal Commission 1952–5*, by the Association of Certified and Corporate Accountants, Day 13, p. 321.

[5] Other illustrations of the use of one-man companies to avoid estate duty are given in Wheatcroft G. S. A., *op. cit.* (1958), pp. 55–7.

professional people avoiding surtax directions;[1] by taxpayers turning investment income into earned income by paying themselves for managing their own investments;[2] and in many other ways.[3]

Among the most important weapons with which the Revenue can counter some of the abuses mentioned here is the power to make a surtax direction against any company which is controlled by five 'persons' or less. Since, for this purpose, two or more near relatives or partners count as one person, and control can be exercised by an aggregate holding of 51 per cent of the shares, it would require considerable ingenuity to remove a one-man company from the purview of this provision. Its effect is that if the Special Commissioners decide that income is being unreasonably retained by the company in order to avoid surtax, they can direct that the whole of the company's income for the year to which the direction relates shall be assessed to surtax as if it had been distributed to the members. The tax so assessed is payable by the company in the first instance. For a surtax direction to be made, someone has to decide what is a 'reasonable distribution'. The onus apparently rests on the Revenue to show 'that the company acted unreasonably in withholding some part of its income from distribution'. It is not enough to show that a part could reasonably be distributed, if at the same time it could be said, as it well might, 'that it was equally reasonable to withhold distribution'.[4]

Two questions arise. First, how many surtax directions are made each year, and how much income is involved? Secondly, how is this income treated in the Board's statistics? Is it allocated to the individual members or not? No information can be obtained from the Board's annual reports and the Royal Commission made no attempt to provide any statistical data on these matters. In any event, for many years during the period we have been studying the Government's policy of dividend limitation made it virtually impossible to invoke these powers in the case of a large number of companies.[5] This meant, as Mr Graham has shown, that 'if a company in 1946 distributed £500 out of a profit of £1,500 it would be free from direc-

---

[1] *Royal Commission, Final Report*, Cmd. 9474, 1955, p. 388, footnote (1).

[2] See Memorandum of Evidence to the Royal Commission 1952–5 by the Association of HM Inspectors of Taxes (*Vols. of Evidence*, Document 32) and *British Tax Review*, March-April 1961, p. 93.

[3] See, in general, *Minutes* and *Volumes of Evidence, Royal Commission*, the works on tax planning cited in this study and Wheatcroft G. S. A., *op. cit.* (1958), pp. 75–6, 134–7, 140 and 161.

[4] Lord Atkin in *Thomas Fattorini (Lancashire) Ltd. v. IRC* (1942) AC 643 at p. 656. For a fuller discussion of the problem see *Royal Commission on Taxation, Final Report*, Cmd. 9474, 1955, pp. 304–18; Wheatcroft G. S. A., *op. cit.* (1958), pp. 134–7, and Graham G. B., *British Tax Review*, March-April 1960, pp. 111–6.

[5] See Memorandum 67 by the Board of Inland Revenue to the Royal Commission 1952–5: 'In present conditions little use is being made of the legislation against "one-man" trading companies.' (p. 658.)

tions in the years from 1949 to 1956 if it regularly distributed £500 per annum out of a profit of £200,000 per annum'.[1] If only a proportion of one-man companies adopted such policies the total effect could be spuriously to deflate a significant number of top income units in the 1949–50 to 1956–7 statistics.[2] As the authors of the Memorandum of Dissent observed, the policy of not enforcing the surtax direction powers 'has enabled successful men in the professions, etc., to form themselves into one-man trading companies and thereby avoid paying surtax on an arbitrarily chosen proportion of their income'.[3] Moreover, by subsequently liquidating such companies it was possible to avoid any surtax charge whatever.

The increasing use of incorporation for (among other purposes) the avoidance of taxation is illustrated by the statistics of new company registrations published annually by the Board of Trade.[4] In 1938 there were 143,221 private companies with share capital registered in Great Britain (excluding those in course of liquidation or removal from the register). By the end of 1960 the number had increased to 363,663. The number of new private companies registered each year settled down after the post-war boom to an average of about 13,000 a year from 1950 to 1953. Then the numbers rose rapidly:

Year	New private companies
1953	13,187
1954	15,703
1955	17,361
1956	17,420
1957	20,531
1958	22,204
1959	28,989
1960	34,058

During the period from the end of 1938 to the end of 1960, the number of *public* companies with share capital on the register fell from 14,355 to 10,806, though their total share capital increased. This trend reflects the growing concentration of corporate power.

[1] Graham G. B., *op. cit.*, p. 113.

[2] The tax problems of such companies are dealt with by a separate 'Companies Division' in the Office of the Special Commissioners. Giving evidence before the Estimates Committee in 1961, senior officials reported that although the Division could not examine in detail every one of 300,000 private companies its work had been increasing. Since 1950, for instance, staff in the executive grades had risen from 24 to about 100 (*Seventh Report from the Estimates Committee*, Session 1960–1, p. 78).

[3] *Final Report*, Cmd. 9474, 1955, p. 388.

[4] Statistics relating to numbers of companies of different types were obtained from the *Companies General Annual Reports* published by the Board of Trade. No information is available on the growth of one-man companies controlled by British subjects but registered in the Bahamas, Bermuda and other countries.

The expansion in the field of private companies, on the other hand, resulted mainly from the registration of large numbers of companies with a very small share capital (the following figures may include public companies, but the number is probably insignificant):

TABLE 6

NUMBER AND PROPORTION OF NEW COMPANIES
WITH NOMINAL CAPITAL UNDER £1,000

Year	No. of new companies with nominal capital under £1,000	% of new companies
1950	4,267	31·0
1951	4,294	32·1
1952	4,446	36·5
1953	5,149	39·0
1954	5,998	38·2
1955	7,189	41·4
1956	7,701	44·2
1957	9,672	47·2
1958	11,079	49·9
1959	15,174	52·4
1960	18,735	55·1

Bearing in mind the effects of inflation these figures show a phenomenal rise in the popularity of the small private company. It might have been supposed that the effect of the Finance Act, 1960, would have been to slow down, if not reverse, this trend, but the figures do not show any such tendency.

We can only surmise that this growth has been stimulated largely by tax considerations. In theory, and given a considerable amount of time, it should be possible to discover something about the objects and activities of a sample of private companies from the documents filed with the Registrar of Companies. In practice, however, this would be a futile exercise. It is true that each company's memorandum of association contains an 'objects clause', but these clauses are usually so widely drawn as to be totally uninformative. The other obvious source of information—the company's accounts—is available only in a minority of cases because nearly 78 per cent of private companies are exempted from filing their accounts with the Registrar, having satisfied certain conditions laid down in the Companies Act. Although this exemption was intended as a protection for the small family business—the genuine 'one-man company'—it seems likely that many other companies which are in fact subsidiaries of non-exempt companies have been able to claim it by having their shares nominally held by individuals. Their affairs are thus removed from the public scrutiny to which the Companies Act was intended to subject them.

The effect of these developments on the Board's statistics of pre-tax income since 1938 is incalculable. Although we know how many companies have been registered, we have no idea how many individuals or families channel all or some of their income through these companies or how and when (if at all) that income appears in the Board's tables of personal incomes. It does, however, seem probable that the use of incorporation for various purposes including tax avoidance has had a significant and increasing bearing on the statistics of income distribution. This possibility does not seem to have been considered in any detail by any students of income distribution since the Second World War, despite the serious attention given to the problem by the Royal Commissions of 1920 and 1955.[1]

We have already noted that Mr Lydall and other income statisticians have commented on 'the relative stagnation' of self-employment and professional incomes between 1938 and 1957 and especially during the post-war period.[2] The shifting of 'income' into retained capital (or undistributed personal savings); the spreading of income-wealth between family and kinship members; and the rise of 154 per cent between 1938 and 1960 in the number of private companies, could explain much of the apparent statistical stagnation in the Board's tables.

Shares for Executives
In recent years a variety of new schemes have been introduced for employee shareholding in industry and commerce.[3] Some take the form of profit-sharing—the schemes run by Imperial Chemical Industries being one example.[4] Others provide bonus shares (or for the purchase of shares at lower than market price) for executive and higher salaried staff.[5] We are not concerned here with employee shareholding in general but only with the capital gains arising from stock option and other schemes, and with any consequential effects such benefits may have in spuriously deflating the earned income statistics.

[1] *Report of the Royal Commission on the Income Tax*, 1920, Cmd. 615, pp. 124–5, and *Final Report*, Cmd. 9474, 1955, pp. 311–6 and 388.

[2] See, for example, Lydall H. F., *op. cit.* (1959), p. 18.

[3] See *The Economist*, November 15, 1958, p. 625, commenting on an 'ingenious scheme' produced by Aims of Industry, and *Review of British Insurance 1961* (*The Economist*, July 22, 1961) explaining the advantages of the 'Executive Savings Scheme' launched by the British Shareholders International Trust in association with the Eagle Star and Guardian Insurance Companies. Savings can be made on estate duty (by policies effected under the Married Women's Property Act) as well as on taxation.

[4] According to the Chairman, Mr S. P. Chambers, unofficial strikers may be excluded from such schemes (*The Times*, May 6, 1960).

[5] For details see *The Times*, January 23, 1959, and February 23, 1960, and an article on share owners by Copeman G., editor of *Business*, in *The Times*, September 22, 1958.

A scheme of this kind, introduced in 1959 and limited to those earning £1,350 or more a year, provides a tax-free capital gain in place of a taxable bonus.[1] Stock options, a common type of executive fringe benefit in the USA, are only taxable in Britain on the value of the option when granted.[2] Any profit made subsequently is thus a non-taxable capital gain. Mr Anthony Vice, writing in *The Director* in February 1962, estimated that fifty major companies had launched share option plans involving £25,000,000 of capital in the previous two years; such plans could 'transform the entire present system of paying directors and senior executives in British industry'. Rewards of this kind, though they may limit the employee's freedom to change his employment, have obvious taxation and estate duty advantages over salary increases. Such rewards are unlikely to be fully reflected in the Board's statistics.

Property and Persons Abroad

'In recent years,' wrote Mr T. D. Scrase in 1957, 'an ever growing number of people have been taking, or contemplating taking, their money out of this country and investing it abroad. While the primary motives for this migration of capital may in many instances have been a desire to secure protection against either or both a devaluation of sterling and any future confiscatory legislation that may be put into effect in this country, there can be few individuals who, having determined to remove their assets, or some part thereof, from the United Kingdom, will not have pondered upon whether by such action they could not also secure some relief from United Kingdom taxation.'[3]

Developments in air travel and other factors in the last fifteen years have revolutionized the possibilities of tax and estate duty avoidance for the residents of many countries. In November 1960, the Chancellor of the Exchequer stated that net British investment abroad during 1950–9 amounted to £2,000,000,000.[4] According to the Board of Inland Revenue, however, overseas interest and dividends paid to British residents in the United Kingdom rose only slowly between 1950–1 and 1957–8.[5] During the balance of payments crisis in the summer of 1961 *The Times* drew attention, in asking the question 'What Has Gone Wrong?', to the lag in overseas interest earnings.[6] So did the Chancellor of the Exchequer in

---

[1] For details see *The Economist*, January 31, 1959. Another company reported in the same month the distribution of £500,000 in shares to forty top executives (*News Chronicle*, January 31, 1959).

[2] *Abbott v. Philbin* (1960) 3 WLR 255; (1960) 2 All ER 763; and see *British Tax Review*, September-October 1960, p. 306.

[3] Scrase T. D., *British Tax Review*, December 1957, p. 356.

[4] *Hansard*, H. of C., November 9, 1960, Vol. 629, Col. 1058.

[5] *BIR 103*, pp. 63–5.

[6] *The Times*, June 12. 1961.

proposing more severe tests for new private investment overseas.[1]

The extent to which the movement and direction in recent years of private investment overseas partly or wholly for avoidance purposes has a bearing on the balance of payments problem is a matter which lies outside the scope of this study. Nor can we go into the questions concerning differential taxation for foreign trading and capital investment overseas and devices to avoid taxation on remittances from abroad in respect of pensions and other sources of income.[2] All we are concerned with are the effects on the personal income distribution statistics of avoidance devices which involve persons and property abroad. These effects have not been studied by income statisticians. The following is one example:

A UK resident buys certain gilt-edged securities which are free of estate duty where the owner is ordinarily resident, and domiciled, outside the United Kingdom. He then settles the stocks on himself for life and thereafter on someone whom he wishes to benefit. His next step is to set up (say in the Bahamas) a company, selling to it his life interest in the securities in exchange for shares. The company, by way of dividend, pays out the income from the gilt-edged stocks to the shareholder. When he dies, however, the company's shares are 'worthless', because its only asset was his life interest which failed at the moment of death. Thus, no estate duty is payable on the shares. Likewise, the trustees of the settlement are exempt from UK estate duty because the beneficial interest in the gilt-edged stocks which comes to an end belonged to a company domiciled abroad. The capital then passes untouched by death duties to the person in whose favour the original settlement was made.[3]

Trustees, donors and testators are advised in many books and articles on tax planning to make provision for powers (or to see that such powers are written into the relevant documents) to buy or otherwise invest in foreign land and immovable property.[4] The law as it affects persons domiciled abroad, in the United Kingdom, or only 'temporarily resident' in the United Kingdom for less than six months in an assessment year, is set out in Professor Wheatcroft's book.[5] There are substantial tax and estate duty advantages in

[1] *The Times*, July 26, 1961.

[2] The taxation aspects were discussed in the *Final Report of the Royal Commission*, Cmd. 9474, 1955, pp. 186–234 and 307. The Board of Inland Revenue memoranda of evidence (Nos. 23 and 51) to the Royal Commission drew attention to the problem of tax avoidance on pensions and other income remitted (or not remitted) from abroad.

[3] Thomson A., City Editor of *The Evening Standard*, February 10, 1960.

[4] See, for example, Potter D. C. and Monroe H. H., *op. cit.*, pp. 397–8 and 405–6.

[5] Wheatcroft G. S. A., *op. cit.* (1958), pp. 101–8, 157–8 and 160–1.

acquiring immovable property abroad (or in transferring certain Government bonds which are free of duty when held by someone resident and domiciled abroad as they can be held by a family company incorporated under another jurisdiction—e.g. the Channel Islands).

These advantages are particularly valuable in such areas as the Isle of Man, Eire, the Channel Islands and countries within the Sterling area. A pamphlet published in 1960 by the Conservative Political Centre points out that Rhodesia, Bermuda, the Isle of Man and the Channel Islands are 'all popular' for investment in land.[1] *How to Pay Less Tax* by Mr A. Thomson of the *Evening Standard* recommended investment in special mortgages in Jersey and Guernsey. He estimated in 1960 that '£1,000,000 invested in this way saves £800,000 in United Kingdom death duties while they stay at their present rates'.[2] Other recommended areas for investment in land included Kenya and Nyasaland.[3]

The Jersey type of mortgage, known as the *hypothèque conventionelle simple*, began to be exploited by tax avoidance money from the United Kingdom around 1954. This traffic rose from £175,000 in 1954 to £3,750,000 in 1960.[4] A Special Correspondent of *The Times* reported that money was flowing in at the rate of £100,000 a week early in 1961, and that the total amount of British tax-avoidance money invested in Jersey mortgages alone had reached £10,000,000.[5] The President of the island's finance committee estimated the potential loss to England in estate duty revenue at £6,000,000.[6]

There are also many interests concerned in offering attractive propositions to British surtax payers in Bermuda and the Bahamas. Real estate agents describe the advantages of acquiring property, land and mortgages in these areas.

There is no direct taxation of any kind in Bermuda and in consequence the residents are free from income tax, death duties, inheritance taxes, and so on. There are no property taxes, other than a small Parish Vestry tax which is quite nominal. Since real estate owned by British subjects and/or residents in countries outside of Great Britain is free of British death duties, this means that any real estate owned in Bermuda is completely free of any form of death duty whatsoever.[7]

[1] Sewill B., *op. cit.*, p. 14.

[2] Thomson A., *op. cit.*

[3] Death duty consultants (some associated with insurance companies) had branches in Kenya, Uganda and Southern Rhodesia.

[4] *The Economist*, February 4, 1961.

[5] *The Times*, June 9, 1961.

[6] *The Times*, February 16, 1961.

[7] Brochures and information available from, for example, Kitson & Co Ltd, Hamilton, Bermuda, and Bahamian Properties Ltd, Nassau, Bahamas.

Easily negotiable 'purchase and lease back' arrangements can mean a saving of £48,000 in British death duties on an investment of £60,000 plus a tax-free capital gain of £15,000 within ten years.[1] By March 1959 land values in some areas of Bermuda had risen to £10,000 per acre. Spacious three-bedroom houses on the waterfront were then selling at £35,000 and over.

In October 1960 the Bahamas International Trust Company Ltd (incorporated in the Bahamas) was advertising the advantages of forming companies and trusts in the Bahamas:

> Companies can be speedily incorporated under the local Companies Act, and there are no corporation taxes. Settlements are administered under the local Trustee Act. Anything that can be done through an English settlement can also be done in the Bahamas, and the Bahamian settlement has certain additional advantages.[2]

A brochure printed by BITCO in October 1959 and distributed by four British banking houses[3] drew attention to the following facts:

> Under Bahamas Trustee law there are no restrictions on accumulations 'thus enabling the Settlor to make long-term provision for dependents'; 'five-man companies' can speedily be incorporated and the Bahamas Companies Act does not require all or any of the directors to be local residents or domiciled in the Bahamas; the law does not require a company to file copies of its accounts or balance sheet with any Government department or agency, nor to declare dividends.

In June 1960, the Isle of Man abolished surtax. The object was, according to one speaker in Tynwald, to bring 'bigger and better tax-dodgers' to the Island. The Chairman of the Manx Income Tax Commission argued that the fiscal policies of the Channel Islands should also be followed by the Isle of Man. He reported that people were going to Jersey in such numbers that they had become a problem and a system of permits had been introduced.[4] According to another

---

[1] Described in detail in brochure published by Kitson & Co Ltd, p. 5 and appendix 2.

[2] *British Tax Review*, September-October 1960, p. ii. Another company, the Union Development Company Ltd, advertised the advantages of buying leasehold property in Nassau and offered to pay 'all reasonable expenses for a purchaser or his representative to visit Nassau' (*The Times*, January 16 and February 23, 1961).

[3] Barclays Bank DCO, Hambros Bank Ltd, the Royal Trust Company of Canada and E. D. Sassoon Banking Co Ltd. Correspondents abroad included companies in Southern Rhodesia and South Africa. Death duty and insurance consultants also name Bermuda and Jersey as attractive propositions.

[4] *The Times*, June 8, 9 and 22, 1960.

report from Jersey in 1960, there had been in the preceding five years 'a rapidly increasing flow of mainland capital, companies and Inland Revenue refugees'. There were said to be 4,000 tax emigrés—twenty of them millionaires.[1]

The formation of family trusts abroad—in countries outside as well as within the Sterling area—is also said to be a profitable undertaking. A typical case, which came before the Court of Appeal in February 1960, concerned the establishment of a trust in the State of Leichtenstein in March 1959 by a British director for the benefit of himself and various relatives.[2] A professor at Oxford University was registered in 1961 as a company in Leichtenstein for taxation purposes, but was said to spend most of his time in Majorca.[3]

Hobby Farming

The problems of avoidance involved in tax-loss farming and other forms of 'hobby' trading were discussed in the *Final Report of the Royal Commission* and by the Chancellor of the Exchequer in May 1960 on certain clauses of the Finance Bill aimed at reducing the loss to the Revenue.[4] Figures produced by the Comptroller and Auditor General in 1959 had shown a loss of over £5,000,000 of tax through charges of farm losses against other income.[5] Losses on farming had been increasing. An estimated 70 per cent of the farmers claiming relief for losses were said, in 1960, to be surtax payers.[6] These losses did not take fully into account the cost to the Revenue and local rates (as a consequence of agricultural derating) of the many personal benefits which can be provided out of farm expenditure; subsidized housing, foodstuffs, gardeners, servants, transport and sporting activities. Horse racing and stud farming is another type of 'hobby' from the point of view of taxation. 'Losses' as a tax deductible are described in detail in the agreement between the Inland Revenue authorities and the Thoroughbred Breeders' Association.

The Ownership of Agricultural Property and Standing Timber

There are many substantial estate duty, taxation and local rate relief

[1] Baistow T., *News Chronicle*, April 6, 1960.

[2] *The Times*, February 26, 1960.

[3] *The Observer*, May 7, 1961.

[4] *Final Report*, Cmd. 9474, 1955, p. 148, and *Hansard*, H. of C., May 24, 1960, Vol. 624, Cols. 317–22.

[5] *The Times*, February 15, 1959.

[6] *Hansard*, H. of C., April 26, 1960, Vol. 622, Col. 6. An article in *The Times* in 1958 drew attention to the advantages of tax relief for retired professional men keeping a 'flock of hens' or doing a little 'market gardening' (*The Times*, December 29, 1958). Under the National Insurance earnings rules for retirement pensions, losses from 'farming' can be set against other income so as to avoid reductions in pensions (Decision No. R(P) 2/60 of the Insurance Commissioner, February 15, 1960).

advantages in the ownership of agricultural land and buildings.[1] According to Professor Wheatcroft, 'woodlands are a particularly attractive proposition'.[2] He also points out that a few days' ownership of agricultural property (technically the 'in and out trick') will enable any donor of a settled gift to reduce liability on his gift by 45 per cent; the ability to fix the date for valuation will enable any number of artificial situations to be created which involve a low valuation of the property in question on that day.

The various opportunities that exist for the conversion of otherwise taxable income into non-taxable capital gains and thence into settled property which may avoid estate duty for up to a century, partly explain the extraordinary rise in the value of farm land and sporting rights in recent years to levels which cannot possibly yield an economic return. An additional reason is to be found in the fact that the larger farms and estates derive more benefit from agricultural subsidies. Insurance companies, death duty and pension consultants, solicitors and accountants are alive to these various considerations.[3] Lord Kilbracken has described the means by which Lord Lonsdale, owning 27,000 acres, has actually increased the real value of his total assets since 1944 despite the payment of well over £2,000,000 in two lots of estate duty. Although it has a very long way to go, the dynasty, maintains Lord Kilbracken, is on the way to owning half the North of England again.[4] 'For the very rich,' wrote the Special Correspondent of *The Times* on the subject of 'Taxmanship in Agriculture', 'the payment of United Kingdom taxation is, to a surprisingly great extent, entirely voluntary.'[5]

Schedule A Imputed Income
Under this heading we are concerned with a particular problem of adjustment which differs in some respects from those we have been discussing in this chapter. The difference lies in the fact that it is not connected with tax and estate duty avoidance. Nevertheless, we need to examine it here because the official classification of income statistics exclude the amount by which the imputed rent of owner-occupied dwellings exceeds the Schedule A valuations. The problem

---

[1] For details and illustrative cases see Lawton P., 'Hobby Trading', *British Tax Review*, July-August 1960, p. 241, and Wheatcroft G. S. A., *op. cit.* (1958), pp. 33–4, 57–8 and 155–6.

[2] Wheatcroft G. S. A., *op. cit.* (1958), p. 156.

[3] For example, see *The Complete Death Duty Manual* compiled for solicitors, accountants, the insurance market and others by Hogg, Robinson and Capel-Cure (Life and Pensions) Ltd, and *How to Pay Less Tax* by Thomson A., City Editor of *The Evening Standard*, 1960. Reference should also be made to Sewill B., *op. cit.*

[4] Lord Kilbracken in *The Evening Standard*, July 4, 1960.

[5] *The Times*, December 12, 1959.

I

of adjusting these statistics arises from the lag between valuations and contemporary property values and, therefore, applies with particular force to the statistics for the 1950's.

Schedule A tax is based on a notional income from a right of occupation which could produce a money income if the owner let his property; by his own choice he enjoys an income in kind of a value equivalent to the income forgone. Compared with those who pay rent for accommodation, equity demands, therefore, that the owner-occupier should pay tax on an imputed income. The question was considered by the Royal Commission on Taxation who endorsed the principle of levying a tax on notional income.[1] These Schedule A assessments are historically designed to represent the annual value as the amount at which the property is 'worth to be let by the year'. For the period covered by this study, however, the notional income was basically related to valuations made in 1934. These were so out of tune with contemporary values, said *The Economist* in 1957, that 'they often represent a mere 1 per cent or 1½ per cent of present-day capital values'.[2] Consequently, this element of tax erosion represented during 1938–57 a growing subsidy from the generality of taxpayers to owner-occupiers—a point made in *Taxes for Today* published by the Conservative Political Centre for the Bow Group.[3] This subsidy was higher for those occupiers living in the larger houses and in the more prosperous parts of the country.[4]

It is difficult to estimate what additions should be made to the Board's personal income data before tax for the difference between imputed income at current values and Schedule A valuations. In 1958 there were approximately 6,000,000 owner-occupied houses in the United Kingdom.[5] If we assume an average notional rent of £60 a year (after deducting repairs and maintenance) the total imputed income for all income units (including those below the exemption limit) would be £360,000,000.[6] The corresponding figures for 1957

[1] *Final Report*, Cmd. 9474, 1955, pp. 245–75.

[2] *The Economist*, February 16, 1957, p. 575. The percentages quoted refer to annual value. Under the Rating and Valuation Act, 1961, householders will be rated in April 1963 on the full current value of their houses instead of on pre-war values, subject to certain transitional derating provisions.

[3] *Taxes for Today*, 1958, p. 25.

[4] The regressive effect of rates was brought out in Mr Lydall's study of household income, rent and rates in *British Incomes and Savings*, Lydall H. F., 1955.

[5] In *Housing Since the Rent Act* the authors estimate the figure for England alone at 5,300,000 in 1958. An addition of 700,000 has been made for Wales, Scotland and N. Ireland (Donnison D. V., Cockburn C., and Corlett T., Occasional Papers on Social Administration, No. 3, 1961, p. 12).

[6] The median annual net rents in 1959 for private unfurnished tenancies in England were reported to be: controlled £38, decontrolled £62. For council tenancies the median was £55 (*ibid.*, table 13). Having regard to the generally higher value of owner-occupied houses the figure of £60 was arbitrarily chosen in

and 1958 estimated by the Central Statistical Office are £223,000,000 and £271,000,000.[1] The Board's tables for Schedule A income do not, however, distinguish owner-occupied houses from all houses, business premises and so forth occupied by persons.[2] It is not possible to say, therefore, how much imputed income has been included by the Board in the tables classifying personal incomes by ranges. Whatever the figure it clearly understates to a considerable extent imputed income at contemporary values. The Chancellor of the Exchequer's estimate in 1960 that it would cost £45,000,000 to abolish Schedule A taxation on owner-occupied houses[3] would, therefore, have to be substantially raised if Schedule A valuations were related to rental values. It is not impossible that the estimate might have to be more than doubled.[4]

Mr Lydall, in his study, did make an adjustment for the excess of imputed rent of owner-occupied houses over their Schedule A valuations.[5] As, however, he used the estimates made by the Central Statistical Office his adjustment did not fully reflect contemporary rental values. Nevertheless, he is the only income statistician who has essayed to 'feed back' into the personal income statistics by ranges some addition for the out-of-date Schedule A valuations.

We make no attempt here to follow him. The only point of this section is to draw attention to this particular deficiency in the official income distribution statistics. What is more important, however, is that it has different effects at different income ranges. The understatement of income applies especially among the higher incomes. It has been estimated, for instance, that the top 1 per cent of income units are assessed for about 20 per cent of aggregate personal Schedule A.[6] While this will include certain categories of properties on which there will have been no tax loss, nevertheless it would seem that the top 1 per cent have benefited disproportionately from the lag in the revaluation of property.

It is also probable that further benefits will have been derived by this group from the more expert and relatively fuller use of allowances for repairs and maintenance, bank overdrafts in respect of

the absence of any comprehensive statistics. For other data on unfurnished lettings see also Ministry of Housing and Local Government, *Rent Act 1957: Report of Inquiry*, Cmd. 1246, 1960.

[1] These are based on estimates of the average rise in rents since the last valuation was made (*NIBB* 1960, table 23 and p. 68).

[2] *BIR 102*, pp. 42–4.

[3] *Hansard*, H. of C., June 21, 1960, Vol. 625, Col. 267.

[4] In discussing Schedule A valuations, *The Economist* employed an index of 2·8 to represent increased values between 1938 and 1959 (January 14, 1961, p. 111).

[5] Lydall H. F., *op. cit.* (1959), p. 26.

[6] *Ibid.*, p. 26.

properties, interest and mortgage payments, the transfer of Schedule A tax from individual to company accounts (a fringe benefit development), the unadjusted imputed rent figure in university 'means test' assessments, and other factors.[1]

In 1959–60 repairs and maintenance relief claims were costing the Revenue £10,000,000, although only 10 per cent of Schedule A taxpayers (6,000,000 houses assessed) made any claim for maintenance relief.[2] A substantial proportion of this sum must have been attributable to the top 1 per cent of income units.

Deduction of Interest and Mortgage Payments
We have already pointed out that the Board's definition of income before tax for the purpose of classifying incomes by ranges is income after the deduction of interest, mortgage payments and other allowable expenses. No statistics have ever been published analysing these deductions by income range, age, sex, marital status and other characteristics. On the assessments made in 1958–9 (all schedules) the 'interest and expenses' deduction amounted to £953,000,000 from a total gross income of £20,300,000,000.[3] This item has grown from £421,000,000 (£11,131,000,000) in 1949–50—more rapidly than the growth in gross income.[4]

Mr E. B. Nortcliffe, writing in the Unilever journal *Progress* in 1959, had this to say about the advantages of living in debt and borrowing for mortgages and other purposes:

> Whilst looking wryly at the way other countries share their taxpayers' burdens, the United Kingdom resident must not overlook those special features of personal income tax in the United Kingdom which enable him to satisfy his penchant for extravagance or speculation at the Chancellor's expense. The saving of both income tax and surtax on interest payments is perhaps the best example. A maximum tax rate of 17s 9d in the £

---

[1] The assessment of parental contributions for grants to university students is an example of the considerable benefits that are derived from the use of statutory income as the basis for the contribution. Those who benefit most are parents in the higher income groups who claim the full allowances for mortgage interest and ground rent on owner-occupied property, interest on bank overdrafts and other interest charges, superannuation contributions and life insurance premiums, domestic assistance, educational expenses for other dependent children, payments made under certain covenants and other items (see Ministry of Education, *Grants to Students*, Cmd. 1051, 1960).

[2] *Hansard*, H. of C., April 28, 1960, Vol. 622, Col. 19, and June 21, 1960, Vol. 625, Col. 267. Additional allowances for Schedule A maintenance relief rose from a total of approximately £20,000,000 in 1949–50 to £55,000,000 in 1959–60.

[3] *BIR 103*, table 29.

[4] *BIR 94*, table 22.

and deductibility of interest together mean that the one thing a rich man can afford is debt. The man of more modest means has less to gain and is in any case usually addicted to hire purchase. From the point of view of tax avoidance hire purchase is distinctly non-U, involving a hire charge curiously misdescribed as 'interest'—a circumstance which certain banks are taking pains to make clear.[1]

Mr Nortcliffe was referring to one of the more striking phenomena of the 1950's; the growth of living in debt among the wealthy—at the taxpayers' expense, a far cry from the Victorian traditions of thrift.[2] This has taken many forms: bank overdrafts; land and property mortgages;[3] the purchase of shares on loan backed by an endowment policy;[4] the purchase of cars by instalments under schemes making use of bank interest payments;[5] and similar devices for the purchase of a wide variety of other expensive consumer goods by people with bank accounts who can, as *The Times* said, offer the sort of security required by the banks.[6]

The development of all these devices in the 1950's for the transformation of consumer goods expenditure and 'hire purchase charges' into bank, building society and finance house interest further enhanced the 'privileged status' (as *The Economist* described it) 'of customers of banks and building societies'.[7] In 1958, when the standard rate of tax was 8s 6d, this privilege was represented by a subsidy of over 40 per cent on interest charges from all taxpayers to those taxpayers living on overdrafts and loans.[8] For those paying interest on hire purchase contracts there was, however, no subsidy. As Lord Kilbracken said, in a letter to *The Times*, for anyone subject

---

[1] Nortcliffe E. B., *Progress*, Spring 1959, p. 79.

[2] Some evidence on this point was provided by the survey 'Savings and Finances of the Upper Income Classes' (Klein L. R. *et al.*, *op. cit.*). Attention was drawn to the 'comparatively large amounts overdrawn in bank accounts by people selected from *Debrett*' (p. 307).

[3] Professor Kahn in his study of *Personal Deductions in the Federal Income Tax* found that the major beneficiaries of the interest deduction in recent years had been homeowners with mortgages. This he described as an 'incentive-subsidy' at the expense of other taxpayers (Kahn H. C., 1960, pp. 123–4).

[4] This development provides another instance of the misleading character of life assurance statistics.

[5] Involving for those paying tax at the standard rate in 1960 a 30 per cent reduction of interest payments. Those with unearned incomes or paying surtax would do better (*British Tax Review*, November-December 1960, p. 379).

[6] *The Times*, October 29, 1958, and October 19, 1960. The proportion of bank customers running an overdraft increased at Lloyds Bank from one in ten in 1957 to one in seven in 1959. At Barclays, the number more than doubled between 1955 and 1959 (*The Economist*, January 28, 1961, p. 375).

[7] *The Economist*, November 1, 1958, p. 436.

[8] *The Times*, October 29, 1958.

to standard tax 'bank interest at even 19 per cent would be a better bargain than h.p. "charges" at 6 per cent'.[1]

The use of interest deductions as a means of helping to pay for private school education (referred to already in the preceding chapter) represents another tax erosion factor which also developed during the 1950's. The 'educational plan', designed by finance (or money-lending) houses in association with insurance companies, has distinct advantages for parents prepared to spread the payment of school fees for five years over a period of ten years. The money-lender pays the school fees, while the parent makes a monthly payment of capital and interest, reclaiming the interest charge as a tax deduction. On a loan of £1,500, for example, to a parent paying tax in 1960 at the standard rate of 7s 9d, the allowance was £15 6s per annum (the gross annual cost of this facility being £43 16s).[2] The generality of taxpayers thus make a contribution of approximately 35 per cent of the interest charges to those parents who take advantage of such plans (apart from the surtax effects). If 500,000 parents on or near the surtax level make use of such facilities the amount of tax forgone could be well over £7,500,000 a year.

Many factors were at work during the 1950's which no doubt helped to foster these practices of technically 'living in debt' among those with sound securities to offer. At certain rates of interest and taxation it paid the wealthy quite handsomely to do so. How much it cost the Exchequer in tax forgone it is impossible to compute. Whatever the cost may be, however, it is clear that, other things being equal, 'living in debt' could exert (what would appear to be) egalitarian effects on the Board's statistics of income distribution.

In the absence of the necessary data we cannot assess the effects of these factors on the presentation of the changing distribution of incomes during the past twenty years by income range, age, sex, marital status, family relationships and so forth. Obviously however, there is an important income differential at work: the combined influence of deductions of this type and their significant growth during the 1950's must have imposed a spurious egalitarian effect on the Board's tables.

Compensation for Loss of Office

Until the Budget of 1960 imposed a limit, these payments were entirely tax-free; they did not therefore enter into the definition of statutory income. Although there were, of course, the genuine cases, the payment of large lump sums masquerading as compensation for loss of office was increasingly used early in the 1950's as a means of avoiding tax. A simple type of case was where a director or senior

---

[1] *The Times*, September 17, 1958.
[2] See, for example, *Educational Plan*, 1960, distributed by Bowmaker Ltd.

employee entered into a contract to serve the company for a period of years at a fixed annual salary and then, quite soon after the execution of the contract and a long time before it was due to expire, agreed to give up his job and waive his claim to his salary in return for a substantial payment.[1]

Such payments, said the Chancellor of the Exchequer in April 1960, 'may represent both a deductible expense for the company that pays them and a tax-free benefit to the recipient. Moreover, as the Royal Commission pointed out, there are startling cases in which, to use their own words, "what is ostensibly a payment in compensation for loss of office is sometimes used merely as a cloak for additional remuneration".'[2] A month later, the Chancellor reported that in recent years 'there have been a growing number of instances where retiring directors, on the occasion of takeover bids, have received very large payments which have become known as "golden handshakes".'[3] In future, only the first £5,000 of any payment was to be exempt from tax. The Royal Commission had recommended in 1955 a maximum of £2,000 for all these cases, in some of which 'contractual rights are created in order to be lost, and nothing is lost which was ever intended to be retained'.[4] The Minority Report had recommended that all such payments should be taxed.

The Government, in rejecting the Royal Commission's recommendation, embodied the limit of £5,000 in the 1960 Finance Act subject to various complicated charging provisions. This limit was not applied retrospectively. Although this tax-free compensation payment may well become in time the normal expectation for departing directors it is clear, as Mr Monroe has pointed out, that the exceptions may be more important than the rule. Under pressure from the House and other quarters, the relevant sections of the Act were 'leniently drafted'.[5] And the penalty on employers for failing to let the Revenue know about compensation payments of, say, £100,000, was fixed at only £50.

In the *British Tax Review* in 1960, Mr Monroe explained and illustrated how more generous payments can still be made—some by disguising them as retirement gratuities.[6] One example given was the payment of a £22,500 gratuity free of tax. Another showed that a compensation payment of £12,000 would, due to the complex ways

---

[1] Board of Inland Revenue Memorandum of Evidence 118, *Vols. of Evidence, Royal Commission on Taxation, 1952–5.*

[2] *Hansard*, H. of C., April 4, 1960, Vol. 621, Col. 55.

[3] *Hansard*, H. of C., May 3, 1960, Vol. 622, Col. 899. Also, 'there is a tendency for these payments to get bigger and bigger as time goes on and also to become more frequent' (May 26, 1960, Vol. 624, Col. 749).

[4] *Final Report*, Cmd. 9474, 1955, p. 78.

[5] *The Times*, April 27, 1960.

[6] Monroe H. H., *British Tax Review*, July-August 1960, pp. 277–84.

in which tax is computed, still leave the recipient with £11,860 free of tax.[1] Moreover, Section 37 of the Act does not apply to any payment made before April 6, 1960, nor to any payment, *whenever made*, being a payment made in pursuance of an obligation incurred before that date.[2]

Compensation for loss of office has increasingly been used and, clearly, will continue to be so used, as a means of converting current and future taxable income into tax-free capital. It is one form of the modern abracadabra which turns everything it touches into capital. How has this development—which all income statisticians have ignored—affected the income distribution statistics, particularly the top 1 per cent of income units?

It is impossible to give any estimate of the total involved in recent years for either genuine or tax avoidance cases. The Government has not published any figures. An article in *The Economist* in 1960 by a Conservative 'who enjoys a professional reputation in the field of taxation' referred to 'the vast sums that have been paid free of tax'.[3] The exclusion of these sums from the income statistics for the 1950's is yet another instance of a spurious egalitarian effect.

These developments during the 1950's meant that the authorities could not collect any tax on lump sum redundancy or severance payments made to manual workers. During this period a growing number of firms adopted such policies. A study published by the Ministry of Labour in 1961 showed that among 236 private companies (employing a total of over 1,100,000 workers and staff employees) 93 made some severance payments.[4] Many of these payments were in cases where only one week's notice of discharge was given. A typical case quoted in this study was a company which graduated the payments according to length of service. The maxi-

---

[1] Other examples given by Professor Wheatcroft in the *British Tax Review* indicated that there 'will now be many cases where "handshakes" substantially in excess of £5,000 will bear little tax' (January-February 1961, pp. 23–29 and March-April 1961, p. 93).

[2] For some years before the Act, various authorities had been advising top executives and directors to enter into agreements to secure 'golden handshakes' in the event of takeover bids or arrangements whereby companies take each other over (see, for example, Thomson A., *How to Pay Less Tax*, 1960, p. 3).

[3] *The Economist*, March 12, 1960, p. 1016. Some of the cases which received mention in the press in 1959 and 1960 included the following: Lord Portal £30,000 and Mr G. Cunliffe £58,000 (British Aluminium take-over, *The Economist*, March 14, 1959, p. 1001); Mr L. Nidditch £40,000 (Ely Brewery and Jasper take-over, *The Times*, June 30, 1960); Mr Perkins £30,000 (Perkins and Co., *New Statesman*, February 7, 1959); Mr Baron £29,237 (Carreras, *New Statesman*, February 7, 1959); Sir Frank Spriggs £75,000 (Hawker Siddeley, *New Statesman*, March 14, 1959); and Lt Col W. H. Kingsmill £60,000 plus annual pension of about £4,000 (Taylor Walker, *The Times*, June 22, 1959).

[4] Ministry of Labour, *Security and Change*, 1961, p. 11 and App. 11.

mum laid down for a worker was £120 after 49 years of service. If such policies were in any way representative of British industry in the 1950's they are unlikely to have affected the income distribution statistics to any significant extent. This could certainly not be said of the sums involved in respect of compensation for loss of office.

Adjustments to Statutory Income

We have so far considered a number of the more important material components of income and wealth which need to be taken into account in measuring the gulf between statutory income and real income. There remain two to be dealt with; the spreading of income-wealth into old age and benefits in kind. These are treated in the following chapters.

The evidence assembled in this and preceding chapters suggests that over the last twenty years or so the gulf between 'income brought under review' and 'accretions of spending power' has been steadily increasing. The causes at work are complex and manifold and appear to arise from the changing characteristics of the social and economic structure of modern societies rather than from any single factor.

The general effect is, however, increasingly to import more and more serious sources of error into the official income distribution statistics insofar as they are interpreted as indicators of income inequalities and not simply as records of 'work done' by the Board of Inland Revenue. Moreover, these sources of error accumulate, so to speak, at the top and affect most heavily the highest income groups. Again and again, in each of the separate categories examined in this chapter, these effects have been observed. The total impact for each of the years surveyed since 1938 is impossible to calculate. The needed information in most instances is not available whether it is sought under the heading of 'income' or 'wealth'.

Two attempts have, however, been made to adjust in some measure the official data for statutory income to take account of certain categories of unallocated personal income: one by Mr Lydall; the other by Mr Brittain. Though it is convenient to refer to these studies here, we shall have to discuss additional adjustments in respect of 'saving' for retirement and benefits in kind in the next two chapters.

Mr Lydall, recognizing some of the defects in the statistics of statutory income made, as he said, a bold attempt to estimate 'the share of the top 1 per cent of the population in unallocated income'.[1] He first considered various sources of income and made separate adjustments under the following headings: life assurance and superannuation, interest on National Savings Certificates, Co-operative Society dividends, the excess of imputed rent of owner-occupied houses over their Schedule A valuation, post-war credits, National

[1] Lydall H. F., *op. cit.* (1959), p. 4.

Insurance contributions, and 'the remainder of unallocated income'. Some of these represent only very minor adjustments. For three of the more important categories (life assurance, superannuation and imputed rents) reasons have been advanced elsewhere in this study to suggest that much more substantial adjustments are needed than those made by Mr Lydall and other income statisticians.

Mr Lydall then assumed (which, as he indicated, was a very questionable assumption) that the unallocated remainder is what it claims to be—the balance of true personal income not yet allocated. On this basis he postulated that the top 1 per cent of income units received 25 per cent of the remainder in 1938, 1949, 1954 and 1957. Without an examination in great detail of the many assorted items in this 'remainder' category, and far more information than is at present available for each of these years, it is virtually impossible to assess the validity of this proportion or the amount of the remainder.

The next step in this series of adjustments that Mr Lydall took was to make additions to allocated income after tax in respect of benefits in kind derived from the social services. These adjustments were limited to health, education, school meals, milk and welfare foods. He assumed that the top 1 per cent received an average share of 1 per cent in 1949, 1954 and 1957 and $\frac{1}{2}$ per cent in 1938. Though these adjustments make very little difference to his aggregate figures after tax they can be criticized on a number of grounds: (1) the limited definition of what is a 'social service'; (2) the fact that insufficient account is taken of the many complicated ways in which the higher income groups may benefit from the high-cost sectors of education, health and other services in kind; and (3) because of the paucity of basic information generally concerning the characteristics of the beneficiaries of many of the public and social services since 1938.

Mr Lydall then essayed to allow for reductions in the purchasing power of income which result from indirect taxation, after taking account of subsidies. Here he concluded that the 'higher income groups probably pay a larger proportion of their allocated incomes in indirect taxes than they did before the war'. But, as he pointed out, there are no adequate sample studies of the expenditure patterns of the top 1 per cent of income units—to say nothing of other units. Consequently, therefore, these adjustments are largely guesswork.

Finally, Mr Lydall made adjustments for undistributed company profits. These have been commented on earlier, and it has been suggested that not only are there considerable deficiencies in the Board of Inland Revenue's basic income figures for the post-war years, but that far more sophisticated analyses are needed to take account of the many complex ways in which the benefits of equity investments may now accrue to families as well as to individuals.

For the result of this courageous and qualified exercise we may quote from Mr Lydall's summary:

> When allowance is made for other sources of income and benefit —namely, unallocated income and social services—as well as for indirect taxation, the fall in the share of post-tax income received by the top 1 per cent of income units is slightly less marked, but still substantial. If, further, allowance is made for imputed income from undistributed profits, which is a procedure of doubtful validity under present conditions, it is found that this adjustment makes less difference to the trend than might have been expected.
>
> Although there are indications that there was a slight tendency towards diminishing inequality before 1938, the rate of change of the past two decades has been exceptionally great. The achievement of full employment has been an important influence; but in recent years the constant upthrust of wages, generated by the wage-price spiral, has been the principal agent changing the shape of the income distribution. For the future, unless there is a catastrophic slump, the trend towards equality is likely to continue, though probably not as fast as in the past twenty years.[1]

Mr Brittain reached quite different conclusions.[2] He started with the original pre-tax statutory distributions, and made adjustments for missing investment income and undistributed company income including stock appreciation, but net of capital appreciation. The result of these exercises showed no levelling of pre-tax income between 1938 and 1949 and 'a remarkable stability in the inequality measure' over the whole period 1938–55.

On the basis of these measures, he then proceeded to consider the effects of fiscal redistribution. Taking the Blue Book incomes after direct taxes, he made adjustments for imputed 'social service' benefits (education, health and housing expenditures and food subsidies), missing investment income, and post-tax undistributed company income. As a result of these calculations, 'the 1949–55 interval shows a 4 per cent rise in inequality, and the over-all 1938–55 decline is more than halved. In sum, the official figures exaggerate the over-all levelling and hide a clear reversal of the trend after 1949'.

Mr Brittain concluded that 'there is no convincing evidence of a "natural" levelling' since 1938. The income 'revolution' which did occur (attributable almost wholly to changes in the top 2 per cent of incomes) 'was a largely inadvertent or accidental by-product of the high taxes, subsidies and dividend restraint required to finance the military budget without runaway inflation' during the war, and since 1949 it has been clearly reversed.

[1] Lydall H. F., *op. cit.* (1959), p. 35.
[2] Brittain J. A., *op. cit.*, pp. 593–603.

The criticisms we have made of Mr Lydall's study are, in many respects, equally applicable to Mr Brittain's study. Their net effects— if further adjustments could in fact be made to both exercises—would be different however. They would weaken the former and strengthen the latter's case.

Two points must, however, be emphasized. Whilst we have summarized in some detail the results of these studies in respect of post-tax income and social service benefits, it should be stressed again that this study is limited to an examination of income distribution *before* tax. Hence we have not explored the effects of direct and indirect taxation and imputed benefits from public and social services. Nor have we examined the effects on the presentation of the income statistics for the lower income units of benefits in kind from these services and concealed subsidies from, for instance, the occupation of council houses. Clearly, these and other factors would have to be assessed in any comprehensive study of equality in the widest terms. But this lies outside the scope of the present book.

Secondly, it should be borne in mind that both Mr Lydall's and Mr Brittain's estimates make little or no allowance for the many important factors of correction and change which we have reviewed and examined in this and earlier chapters: the serious defects and discrepancies in the basic data; the size of the tax universe and its division into parts; changes in age, sex and social structure; the definition of the income unit; the effects, through a variety of complex mechanisms, of spreading, splitting and transforming income-wealth over time and amongst members of the wider kin group, and other factors.

One of the consequences of the growth and proliferation of these factors is a steady and continuing erosion of the tax base; what Mr Kaldor has described as 'the path of charging more and more on less and less'.[1] Perhaps, however, this over-simplifies what has actually been happening to fiscal policies during the last decade. While the tax base—and especially the surtax base—has certainly been narrowing rapidly, the process of 'charging more and more' has been directed towards flat-rate and regressive systems—notably National Insurance—rather than direct and progressive taxation. The growing weakness of one system of taxation has led to a search for another system not subject to this process of erosion. We return to this theme in our final chapter.

Meanwhile, we should draw attention to two other important developments which affect this process of erosion. One is the growth of social policies in fiscal law during the past fifty years.[2] These

[1] Kaldor N., *An Expenditure Tax*, 1955, p. 242.
[2] For some account of this trend see Titmuss Richard M., *Essays on the 'Welfare State'*, 1957.

policies have increasingly recognized the dependencies of the individual and the family in a greater variety of situations. The result has been a steady enlargement in the value and range of personal allowances—for children, parents and other relatives, child-minding, further education, old age, wives at work, housekeepers and so forth. These developments do not, of course, affect the primary statistics of statutory income—the income reviewed before the deduction of personal allowances. Hence, they do not immediately concern us here. In general, however, it may be said that the growing importance of these social policies in fiscal law has been neglected by most income statisticians who have drawn conclusions from studies of income *after* tax.

Another development contributing to this process of erosion in the tax base has taken the form of disregarding for taxation purposes a greater variety of monetary receipts. They fall into no single category at all—apart from their legal recognition as 'disregards'—for they range from football pool winnings to tax-free expenses for members of the House of Lords. To conclude this chapter we append a representative list. None of the items enumerated is included in the Inland Revenue statistics of income before tax. What the total involved amounts to it is impossible to estimate without a great deal of laborious research (in a few cases where the figures were ascertainable they have been added). Nor can we estimate in what manner these developments have affected the statistics of personal income before tax and by ranges since 1938, and what further adjustments should be made to the statutory income distributions.

The following allowances, payments, receipts and benefits were tax-free at the end of 1960:

1. *The Armed Forces.* Allowances, terminal grants, bounties, gratuities and benefits in kind; training expenses allowances and training bounties; allowances and bounties to reserve and auxiliary forces; re-engagement gratuities; Korea gratuity; Victoria Cross annuities and additional pensions; war wound and disability pensions; war pensions to widows in respect of their children; educational allowances for children at boarding schools for members of the Armed Forces serving overseas. Scholarship awards to Dartmouth, Sandhurst and Cranwell.

The Treasury estimated that the total amount of payments in 1959–60 exempted from income tax under Section 457 of the Income Tax Act, 1952, was about £30,000,000.[1] Since then substantially higher gratuities have been approved. Thus, officers in the Fleet Air Arm now receive a tax-free gratuity of £4,000 after twelve years.

---

[1] *Hansard*, H. of C., February 17, 1961, Vol. 634, Col. 189. Section 457 specifies the gratuities, bounties and allowances of members of the Armed Forces, etc., not regarded as income for any income tax purposes.

Doctors in the Army receive £3,000 tax-free after five years' service.

2. Civil servants' foreign service allowances. Pensions payable under the Superannuation Act, 1949, in respect of a civil servant's children are regarded as income of the children for tax purposes. Civil servants' death benefits under the Superannuation Act, 1909, are not subject to estate duty. (Tax-free retirement benefits for civil servants are referred to in the next chapter.)

3. Expense allowances for members of the House of Lords.

4. Post-cessation receipts for barristers.

5. Compensation to individuals and companies under schemes for the reorganization of the cotton industry under the Cotton Industry Act, 1959. An amount of approximately £30,000,000 was said to be involved.

6. Some part (according to the tax position of the individual) of compensation payable under the Town and Country Planning Acts, 1954, in respect of the depreciation of land values, refusal of planning permission and so forth. From January 1955 to March 1961 the total compensation paid was approximately £98,000,000 (excluding interest).[1]

7. Non-trading transactions. Isolated transactions or the purchase and sale of an asset will not give rise to an income tax assessment unless it is 'an adventure in the nature of trade'. The profit arising on the sale of an investment (or of an asset held for the income or enjoyment it produces rather than simply for the purpose of resale) is not taxable. Stockbrokers therefore are not assessed on their dealings, since a broker (unlike a jobber) is not in business as a share dealer. If he acquires shares he is assumed to do so for investment purposes. This is true even if he sells them before he buys them.[2]

8. Arrears of dividend on preference shares paid on the liquidation of a company are regarded as a distribution of capital and are not taxed.

9. Receipts from the sale of goodwill.

10. Interest on National Savings Certificates. It has been estimated by Mr Lydall that the top 1 per cent of income units owned about 25 per cent of all Certificates in 1957–8.[3] Mr Enoch Powell, after analysing the individual forms of National Savings over the last ten years or so, concluded that these Savings 'have now remarkably little connection with saving'.[4]

11. Prizes paid on Premium Bonds and interest on Tax Reserve Certificates.

12. The first £15 of interest from post office, seamen's and trustee

---

[1] *Hansard*, H. of C., April 19, 1961, Vol. 638, Col. 113.
[2] See note on 1962 budget, p. 143.
[3] Lydall H. F., *op. cit.* (1959), p. 26.
[4] Powell J. Enoch, *Savings in a Free Society*, 1960, p. 93.

savings banks is exempt from income tax, but must be grossed up for surtax. Income tax forgone on account of this exemption amounted to about £12,000,000 in 1959–60.[1]

13. Income arising from a scholarship or similar endowment held by a person receiving full-time instruction at a university, college, school, etc., is exempt from income tax and surtax. Such income is also ignored in determining whether the parents are eligible for Child Allowance. The Exchequer was unable in 1961 to estimate the amount of tax forgone.[2]

14. National Insurance sickness, unemployment, maternity and death benefits.

15. Sickness and disablement benefits from mutual insurance societies, if for periods of less than a year.

16. Concessional allowances to miners in lieu of free coal.

17. Damages awarded for wrongful dismissal.

18. Certain redundancy payments for discharged workers.

19. Betting, gambling and football pool winnings.

[1] *Hansard*, H. of C., February 17, 1961, Vol. 634, Col. 189.
[2] *Ibid.*, Col. 191.

Note on 1962 budget.
    The position regarding the taxation of speculative gains is affected by the 1962 budget. The proposals for taxing certain types of short-term transactions (not, as the Chancellor emphasized, a capital gains tax) were, according to *The Times*, 'well received in the City' (*The Times*, April 10, 1962).

CHAPTER 7

The Spreading of Income over Retirement

I

IT is not the purpose of this chapter to survey developments in various forms of 'savings' for retirement and 'deferred pay' since 1938. The subject is one of extraordinary complexity which will, it is hoped, be comprehensively treated in a separate and later study. The approach adopted here is severely restricted both in scale and in time. We consider only the possible effects on the income distribution statistics since 1938 of changes in these categories of savings or deferred consumption or methods by which income is transformed into capital. The many factors discussed in earlier chapters and particularly the striking changes in the social and demographic structure of the population make it important, however, to deal with the question in some detail.

Another reason for doing so lies in the fact that statutory income— or pre-tax income brought under review by the Board of Inland Revenue—excludes National Insurance and superannuation contributions by employees and employers. There is no general agreement as to whether these contributions and the net increase in life and pension funds (the collective property of the members) should be treated as personal income. Much depends on why the question is being asked. A photographic approach to the distribution of incomes by ranges at a given point in time may not regard these contributions (or savings, deferred consumption or increments to future spending power) as personal disposable income. Such an approach would be supported by the element of compulsion or taxation inherent in many pension and superannuation schemes both private and public. In any event, there is the real statistical difficulty of allocating employers' contributions and the collective income of funds to personal income ranges.

In this study, however, we have not in general been taking a photographic view of the distribution of incomes. On the contrary, we have, for particular purposes, argued the case for a longer view.

Changes in the structure and organization of modern society make this increasingly necessary. An approach to the measurement of inequalities in income-wealth which ignores the many ways by which the constituents of spending power—present and future—may be spread, split and transformed from one constituent to another is becoming more unreal. The evidence collected in Chapter 5 in particular leads to this conclusion.

For some purposes, of course, a 'snapshot' view of income distribution is useful. If the statistics were available in terms of age, sex and family relationships we should, for instance, learn something about the immediate effects of arrangements for spreading income-wealth among the members of the family. These effects might then be interpreted in terms of the individual head of the family as a spurious equalitarian effect, although they may be differently considered in terms of the family as a total unit. When, however, we examine those arrangements which spread income-wealth over the life of an individual or over several family generations a longer view is needed. Ideally, of course, both views are required if we wish to be better informed about the changing distribution of income and wealth. It is the absence of statistics 'in depth' and the inadequacies of existing 'snapshot' data which have led in large measure to the criticisms we have made of statutory income in this study.

For these reasons, therefore, we propose to treat all contributions, increments to pension funds, and life assurance premiums as a part of personal disposable income before tax.[1] In short, the capacity to save—to store up for retirement—is part of current standards of living. If it is partly or wholly done by one's employer or some other agency the same principle holds. In order to obtain, therefore, a true picture of pre-tax income, adjustments should be made to the Board of Inland Revenue data since 1938, and account take of developments over the same period in this field of saving for retirement and other purposes.

We consider the problems involved in making these adjustments to the Board's figures after we have briefly reviewed the growth of superannuation and pension schemes since before the War. First, however, it is necessary to draw attention to one of the oddities about these income statistics. While superannuation contributions

[1] There is general agreement that National Insurance contributions are a tax. They are so treated by the Central Statistical Office, by Mr Lydall (*British Incomes and Savings*, 1955), by Professor Peacock (*The Economics of National Insurance*, 1952), by Mr Enoch Powell, who also includes Health Service contributions (*Saving in a Free Society*, 1960), and other authorities. The present writer accepts this classification. To regard these contributions as taxes, however, does not affect the analysis in this study which is concerned with the distribution of pre-tax income.

K

by employees are deducted in arriving at personal income before tax, life and endowment assurance premiums are not. Yet the latter are often effected for a similar purpose: the spreading of income into old age for the 'income unit' of husband and wife.[1] Moreover, for many directors, self-employed business men, members of certain professions and others in the higher income groups, considerations of estate duty saving have led, in recent years, to a closer relationship between life cover, retirement provisions, tax-free capital sums and benefits for wives, widows and other members of the family.[2] A brief historical account of the development of this relationship is given by Mr Tony Lynes in Appendices D and E.

---

[1] This is reflected in the fact that some superannuation contributions are treated as life assurance premiums (e.g. the Federated Superannuation System for Universities) which adds another confusing item to the statistical picture. Writing of life assurance relief and superannuation contributions, the Board of Inland Revenue in a memorandum of evidence to the Royal Commission had this to say: 'Broadly speaking, these two reliefs can be regarded as methods of alleviating the burden of tax upon precarious incomes where the recipient has somehow or other to try to provide for the future and as such they can be regarded as incidents of graduation though they are also inducements to save' (Memorandum 111, *Vols. of Evidence, Royal Commission on Taxation, 1952–5*).

[2] This applies particularly to benefits for widows and other dependents. Many retirement schemes and trust deeds for salaried directors and higher paid staff make such provision. As Messrs Potter and Monroe point out: 'It is usually more advantageous if in the case of a pension to a widow (or other dependant) the contract or scheme provides for two separate pensions, one to the retired employee, and, after his death, a separate pension to his widow (or other dependant) and not one pension which is enjoyed in succession. The decision on this point is *Re Payton* (1951) Ch. 1081 (see also *Re Weigall's Will Trusts* (1956) Ch. 424). In this case, the Austin Motor Co Ltd entered into a contract with an insurance company for the purpose of providing pension benefits to employees. An employee had an option (which was treated as exercised) to surrender the pension to which he was entitled for "a last survivor pension" which consisted of a smaller pension to himself and, if his wife survived him, to his wife for the rest of her life. It was held that as there were two separate interests, there was no passing on the husband's death and that estate duty was payable under section 2 (1) (d), and not section 1, of the Finance Act 1894. As a consequence the benefit to the widow was exempt from aggregation with the rest of the deceased's estate because the deceased never had an interest in the property deemed to pass.' (Potter D. C. and Monroe H. H., *Tax Planning*, 1959, pp. 323 and 332–3). In their study of 'The Growth of Life Assurance in the United Kingdom since 1880', Messrs Johnston and Murphy illustrate 'the immense attraction' of separate policies written under the provisions of the Married Women's Property Act, 1882. Appendix E shows something of the flourishing nature of avoidance and evasion devices developed by insurance companies after 1943 which made use of the 1882 Act. No information at all has been published by insurance companies of the amount of business transacted since 1938 under these provisions. It may be that a large part of the 'life cover' for employees effected by employers and paid for by employers since 1938 falls into this category (Johnston J. and Murphy G. W., *The Manchester School of Economic and Social Studies*, 1957, Vol. XXV, No. 2, p. 107).

During the past fifteen years or so a variety of intricate schemes have been designed which combine some or all of these benefits in addition to effecting a saving of both income tax and estate duty. Illustrations have already been given earlier in this study of some of the ways in which changes are now made in the time and kinship pattern of consumption. The pension consultant (like other taxation consultants) has become interested not only in such instruments as discretionary trusts, but also in such problems as 'what constitutes a family' in fiscal law and administrative practice.[1] As these schemes become increasingly complex with the merging together of the three major objectives of 'spreading income over life', 'splitting income between the kinship', and 'converting income into capital', it is virtually impossible to distinguish the 'pure' life assurance element from other elements, particularly when premiums are paid by employers in consideration of acceptance of a lower salary. At the higher levels of income, therefore, the different statistical treatment accorded to these elements by the Board of Inland Revenue adds one more confusing factor to those previously described.

To be consistent (and quite apart from the actual taxation position) both superannuation and life assurance payments by the taxpayer should therefore be treated in the same way in the relevant pre-tax tables. If the purpose of these statistics is to depict the distribution of incomes before tax, all such payments should not be deducted. They are forms of expenditure or of allocating disposable income, and there is no justification for treating them in any other way.[2] The fact that the benefits are taxed when they are received is not relevant to the purposes in hand of measuring the distribution of incomes before tax and presenting the data in a consistent form. This is part of the general problem we have met before; of distinguishing between taxable income and income for the purposes of measuring inequality.

Because of the way in which the official statistics are at present arranged it thus becomes necessary—if we wish to study the changing distribution of incomes—to feed back all contributions and payments by individuals. Those made by employers (actual or notional as in the Civil Service and Armed Forces) should be similarly treated; so also should all life assurance and endowment premiums paid by employers for their employees and members of their employees'

---

[1] See, for example, *Pension Schemes* (1960) by two insurance experts, Messrs M. Pilch and V. Wood, especially the section on 'Controlling Directors and Family Control'.

[2] As National Insurance and Health Service contributions are compulsory for the majority of people and form part of the present system of direct taxation they should logically be represented as such in the income *after* tax tables (see *National Income Statistics: Sources and Methods*, 1956, pp. 72 and 190).

families.[1] Moreover, allowance should be made for the fact that in a large and growing number of deferred pay arrangements a substantial part of the eventual benefit may be taken in the form of a tax-free lump sum. In other words, in return for a reduction in disposable income before tax a tax-free capital gain accrues in a given year. The individual is rewarded by the generality of taxpayers for altering the time pattern of his consumption.

II

While it is comparatively easy to lay down the principles on which the income distribution statistics should be arranged it is quite another matter to make all the necessary adjustments. As we shall show, the difficulties are extremely formidable, and many of them derive from developments in provisions made by employers.

Contributory and non-contributory superannuation payments and life assurance premiums paid for by employers as a business expense rise very sharply with income. This is particularly true of non-contributory superannuation schemes which, over the past fifteen years or so, have become increasingly common among employees in the higher income ranges. This was not so before 1939. According to Messrs Pilch and Wood, group life and pension schemes were 'in the early years anyway, almost always on a contributory basis.'[2] The 1937–8 income statistics were, therefore, differently affected from those in 1949–50 and subsequently. Hence they call for different adjustments in the process of feeding back contributions and payments.

As a result of the development since 1938 of 'top hat'[3] and a variety of complex individual schemes, an increasingly large slice of current income has been spread into retirement. Some account of this development is given in Appendix E. The origin of these schemes in the 1940's is said to have been the recognition of the value of exchanging a lower salary for retirement and life cover benefits plus a tax-free capital sum in the sixties.[4] Just after the war, a director aged forty-five could relinquish £5,000 a year of salary at a net cost

[1] In the National Income accounts employers' contributions and the property and investment income of the funds of superannuation schemes and life offices are treated as part of personal income (*ibid.*, p. 67).

[2] Pilch M. and Wood V., *op. cit.*, p. 22. A similar conclusion was reached by Messrs B. Robertson and H. Samuels: '. . . the principle of contributory as against non-contributory schemes is so firmly established that we can . . . confine ourselves to . . . funds membership of which involves some contribution on the part of the employee' (*Pension and Superannuation Funds*, 1930, p. 7).

[3] According to Messrs Pilch and Wood, the sobriquet was apparently invented by a writer in the insurance press with the intention of disparaging what he regarded simply as a device for tax evasion, unworthy to occupy the time and attention of insurance companies.

[4] Pilch M. and Wood V., *op. cit.*, p. 26.

of only £250 a year in spendable income. In exchange for this he would receive 'a sum assured of at least £100,000 payable at age sixty-five or previous death, and the whole of this amount could be paid on retirement as a tax-free capital sum'.[1] All endowment assurance schemes of this kind and established before 1947 may give their members the right to take the whole of their benefits as a tax-free capital sum on retirement.[2] Some such schemes which have matured in recent years or are in existence today may provide benefits of the order of £250,000 or more. Moreover, the 1947 legislation did not affect new entrants after 1947 to schemes already in existence before that year. Tax-free lump sums of this magnitude may still be given to such entrants.

Discussing, in 1960, the extent to which 'top-hat' pension schemes were creating a privileged group, *The Economist* reported that such schemes had 'grown to great dimensions' in recent years.[3] The Board, it seems, has no record of how widespread these schemes are and how many tax-free capital sums exceeding £100,000 have been paid in the last ten years or may be paid in the future. The Millard

[1] Pilch M. and Wood V., *op. cit.*, p. 28.

[2] The Millard Tucker Report described the situation before 1947 in the following terms:

'It has already been explained in para. 52 that before 1947 an employee was taxable in respect of payments made by his employer to secure future benefits for him after the cessation of his service only in the very exceptional type of case in which his interest in each such payment was absolute. He was immune from such taxation if his interest was contingent, for example, if his title to receive the benefits depended upon his remaining in the employer's service until a specified age, or on an ultimate discretion exercisable by the employer. Accordingly, it could be arranged, and often was arranged, that, although all the money laid out by the employer in relation to the employee's service ranked as a deductible trading expense of the employer, the employee during his actual service was not liable to income tax and surtax on that money, but only on the current remuneration he actually received. The amounts contributed by the employer towards future benefits accumulated until the benefits fell due, and at that stage the employee became liab'e to tax, but then only in so far as the benefits were in pension form. Alternatively, instead of making actual payments to secure the future benefits, the employer could simply agree, as one of the terms of the employment, to pay such future benefits himself when the time of retirement came. Those payments would then be treated as part of his current trading expenses. All this opened the way to devices for giving what were clearly excessive benefits on the cessation of service, and giving them at the expense of current taxable remuneration, so reducing the employee's current liability to both income tax and surtax. Moreover—and this is perhaps even more important—the ultimate benefit on cessation of service could be payable in the form of a single lump sum, or in the form of a pension commutable at the option of the recipient for a single lump sum, and in either case the lump sum escaped tax altogether.'

The Report went on to provide illustrative material all of which has escaped the notice of those who have studied the distribution of incomes (Cmd. 9063, p. 25, *et seq*).

[3] *The Economist*, March 12, 1960, p. 1017.

Tucker Report said that there were 'some 27,000 non-statutory staff funds and schemes' in operation in 1953, 'and also a large number of schemes for selected individual employees. The number of employees concerned in [all] these funds and schemes is not known, but a rough estimate puts the figure at about 6,000,000.'[1] Statistics published in the *Policy Holder* in 1951 suggested that 'a figure in the region of £75,000,000 would be required to cover 1950's "Top Hat schemes", and that the 1951 total considerably exceeds that figure . . . they are often written on lives of a fairly advanced age.'[2] Without much more detailed information, however, no precise adjustments are possible to the income distribution statistics. Clearly, if they could be made on this score of lump sums alone the statistical picture for the top income groups for recent years would look very different from that for 1937–8.

The studies of Messrs Pilch and Wood and other insurance and legal authorities show the extent and bewildering variety of schemes in existence in 1961 which make provision for tax-free lump sums on retirement or death of up to a quarter and, under some schemes, of the whole of the benefit.[3] The effect on the individual 'income unit' in the Board's statistics is apparent from the following illustration. It is taken from a typical and widely distributed advertisement in the 1950's:

	Case 1			*Case 2*	
An employee, aged 41, married with three children, earns		£5,000	An employee, aged 41, married with three children, earns		£4,300
Having no pension rights, he has taken out an endowment policy for £1,750 annuity at sixty-five. To pay the premium of £750 he has to set aside from salary		£1,780	His firm provides its executives with 'Top Hat' pension policies. His pension will be £1,750. The premium is £700 but it costs him		Nil
The tax on the remainder of his salary amounts to		£1,120	His income tax amounts to		£1,800
Thus he has left for all normal expenses		£2,100	Thus he has left for all normal expenses		£2,500

'The "Top Hat" man has £400 a year more spending money, although he will retire on *exactly the same pension* and earns a smaller salary.'[4]

The tax advantages are still greater if part of the retirement benefits are taken in lump sum form. Thus, at 1959–60 rates of tax, a director retiring at sixty-five with a 'top hat' pension of £11,500 a

[1] Cmd. 9063, p. 79.
[2] *Policy Holder*, December 5, 1951, Vol. LXIX, p. 1230.
[3] Pilch M. and Wood V., *op. cit.*, pp. 29, 35 and 63–4.
[4] *The Times*, January 21, 1954.

year and a lump sum of £40,000, and surviving for twelve years after retirement, would have paid nearly £40,000 less in income tax and surtax than if the whole of the benefits took the form of a pension. To achieve this result, his salary would have been 'reduced' from £30,000 a year to £23,100.

Many permutations on this and similar themes have been played during the last fifteen years.[1] Some have included additional and separate death benefits provided by the employer for his employees and their dependents under retirement benefit schemes; developments in these fields and in respect of tax-free lump sums are reviewed by Mr Tony Lynes in Appendices D and E which illustrate the role played by insurance companies in the provision of such schemes. Others have taken the form of a deliberate reduction of salary in return for non-contributory pensions, tax-free lump sums and other benefits provided by the employer. Many others have meant forgoing salary increments. The benefits have included a variety of provisions; some, which have grown in popularity in recent years (though generally restricted to surtax payers) constitute instalment or loan schemes for share purchase under endowment policies. The advantages of these 'accumulation unit' funds are that the management expenses are allowable for tax purposes, the premiums paid in are fully allowable for tax purposes, and the dividend income of the funds is tax-free.[2]

All these developments involving the manipulation of salaries and fees in relation to estate duty and tax advantages may help to explain what many income statisticians have regarded as a puzzling characteristic of post-war income statistics. Mr Lydall, for example, commented several times on 'the persistent failure of pre-tax allocated incomes in the upper ranges to rise as fast as pre-tax allocated incomes in the middle of the distribution'.[3] On the face of it, therefore, a trend towards greater equality of incomes before tax is observed. But as no studies have been made of the extent to which

---

[1] Mr C. R. Lowndes (a well-known authority in the life insurance world) reported in 1959: 'it is extremely rare to find a company of any size that does not operate [a Top Hat Scheme] for its senior staff at managerial level' (*The Spectator*, November 27, 1959, p. 796).

Mr. Copeman, in his survey of executives in 1957, said that a 'large number of senior men were found to be covered by a Top Hat pension policy in addition to their company's normal scheme for staff. One director said that in his company all those earning £3,000 a year or more are allowed to join the Top Hat scheme and forgo some of their highly-taxed salary in order to increase future pensions. Another director said that he has 'assigned' £2,000 of his salary to the Top Hat Scheme.' (Copeman G., *Promotion and Pay for Executives*, 1957, pp. 199–200).

[2] *The Times*, March 9, 1959, June 5, 1959 and January 12, 1960 and *The Economist*, June 13, 1959 and July 22, 1961.

[3] Lydall H. F., *op. cit.* (1959), p. 8.

high salaries were reduced or increments were forgone between the early years of the Second World War and the middle 1950's it is not possible to hazard a guess as to the effects on the income distribution statistics.

III

There is a good deal of evidence to suggest, however, that for the middle and upper-middle classes the proportion of those with non-contributory benefits of various kinds rises with income. The Royal Commission on Doctors' and Dentists' Remuneration showed, for example, that 71 per cent of actuaries in salaried employment (the highest paid group among all the non-medical professions considered) had non-contributory benefits.[1] This group far outdistanced the next on the list; ironically, it is this group of professional men whose traditional role has been to preach the economic and moral virtues of sound contributory insurance schemes. In respect of non-contributory retirement benefits (considered alone) they are probably closer to civil servants than any other single category of employees.

The survey by the Government Actuary in 1957 of occupational pension schemes showed:[2]

	Private Schemes	
	Insured	Non-insured
Proportion of schemes with no employee contribution	30%	44%
Proportion of cases in which partial conversion of pension to lump sum is permitted	53%	43%

Regrettably, this Report provided no breakdown of the figures by age, sex, range of benefit or income group.[3] Such evidence as does exist, however, points to the growing importance among higher income groups (especially for younger men) of the three factors of non-contributory pensions, life cover and lump sum benefits. A survey of over 700 executives by the magazine *Business* in 1960 showed that one-third of the under-40 age group were insured by their firms for over £10,000.[4]

So far as retirement provisions and tax-free lump sums amounting to at least one-quarter of the benefits are concerned, it would seem from the evidence that by 1961 a large proportion of all higher paid

[1] Cmd. 939, 1960, p. 43. The highest proportion of those earning £10,000 or more were also actuaries.

[2] *Occupational Pension Schemes*, A Survey by the Government Actuary, 1958, pp. 18 and 24.

[3] 'It has been found unnecessary,' said the Report without further explanation, 'to show variations in the provisions for the sexes except as regards retirement ages' (*ibid.*, p. 3).

[4] *Business*, August 1960, p. 47.

salaried employees were generally in the same position as civil servants in the administrative class.

From the Millard Tucker Report in 1954 and the Report of the Royal Commission on the Civil Service 1953–5 it appears that the cost of Civil Service pensions at that time amounted to approximately 18 per cent of actual salaries.[1] This figure excludes the cost of widows' and dependents' benefits. It also excludes the loss of tax through giving lump sum benefits and the loss resulting from spreading income over retirement.[2] The value of these lump sum benefits has risen substantially in recent years as the following figures illustrate:[3]

TABLE 7

LUMP SUM BENEFITS IN THE CIVIL SERVICE

After 40 years' service	1939 Scale		1961 Scale[4]		Increase in tax forgone	
	Lump sum free of tax	Tax forgone	Lump sum free of tax	Tax forgone		
	£	£	£	£	£	%
Permanent Secretary	4,500	2,035	10,092	8,163	6,128	301
Assistant Secretary	2,250	763	4,815	2,949	2,186	287
Principal	1,650	476	3,498	1,838	1,362	286
Clerical Officer	525	109	1,182	356	247	227

Difference in the value of tax-free lump sums between Permanent Secretary and Clerical Officer

	1939 Scale	1961 Scale	Increase	
	£3,975	£8,910	£4,935	124%

Difference in the amount of tax forgone between Permanent Secretary and Clerical Officer

	1939 Scale	1961 Scale	Increase	
	£1,926	£7,807	£5,881	305%

For both years the hypothetical tax was calculated on the assumption that the retiring officer was a married man without dependent children; that retirement took place on April 5th so that the lump sum was payable in addition to a full year's salary; that there was no other income; and that the lump sum was taxable as earned income. The assumptions that the whole lump sum is taxed in one year and that a full year's salary is received in addition are generous ones from the point of view of the Exchequer; however, the figures serve to show the maximum amounts of tax forgone.

[1] Cmd. 9613, 1955, and Cmd. 9063, 1954, p. 41.

[2] In the private sector the provision of generous back-service rights by employers for higher paid employees could well bring the cost to 25 per cent or more of gross salaries, apart from the costs to the Exchequer and the employer of tax-free lump sums, life cover, widows' benefits, increases in benefits after retirement, and tax forgone through spreading income over retirement (an estimate of '25 per cent or more' was given by Mr E. Innes in the discussion on Mr Lydall's paper at the Royal Statistical Society, op. cit., p. 43).

[3] Hansard, H. of C., February 17, 1961, Vol. 634, Col. 191.

[4] Assuming 1959–60 rates of surtax.

If accurate figures were available for the whole of the private as well as the public sector of employment the total loss of tax, as a result of this unique British provision of tax-free lump sums, must have assumed by 1960 very high proportions.

In summarizing these developments in superannuation, lump sum and life assurance schemes since 1938 we may put forward certain propositions. They are stated here as a preliminary to tracing the growth over the last twenty years in the numbers covered by occupational pension schemes.

1. That in absolute and relative values of tax-free lump sum benefits the higher paid civil servants have gained more than the lower paid. Lump sum and pension benefits are greater in relation to pre-retirement earnings for the higher paid than for the lower paid. In these terms, therefore, inequalities in the civil service have increased (partly due to the close relationship between these benefits, promotion in the last ten years of service, and peak earnings).[1] To a large extent, the nationalized industries and other public services have adopted or followed the civil service model. For higher paid executives and salaried officials in the private sector, the civil service system of benefits had by 1947 become the *minimum* model. In an unknown proportion of cases the lump sum, life cover, widows' benefits and other provisions are far more generous.

What the adoption of this model may mean in terms of retirement benefits alone for higher paid employees in the private sector was illustrated in 1961 by an appointment to one of the nationalized boards. The salary of the person appointed was the same as he had previously received, namely, £24,000. It may be calculated that a civil servant with a final salary of £24,000, retiring at sixty-five after forty years' service, would receive a pension of £12,000 a year and a tax-free lump sum of £36,000. The capital sum required to purchase an annuity whose capitalized value *after* tax (at 1960–1 rates) would be £36,000 is £314,155, assuming that the employee is married with no other allowances and no other income except a National Insurance pension of £208. The amount of tax forgone would indeed be considerable.

2. That the proportion of tax forgone by the Exchequer in respect of lump sum benefits rises with salary, and that since 1938 these differentials have become longer and more pronounced (partly due

---

[1] A study by the International Labour Office in 1953 examined the salary and wage differentials within civil service systems between selected higher grades (e.g. heads of specialist services and junior administrative officers) and lower grades (e.g. letter carriers and messengers). The differences were considerably greater in the UK than in the USA, Canada, Denmark, Norway, Australia and other countries (*Salaries and Hours of Work in Government Service: An International Comparison*, International Labour Office, 1953).

to the factors already mentioned and partly to the lengthening in the bureaucratic hierarchy).

3. That over the last twenty years the retirement benefits of more and more salaried white-collar workers in public and private employment have approximated closer to those provided for civil servants.[1]

4. That the great expansion in the numbers covered and the value of retirement benefits in the private sector between 1938 and 1960 took place chiefly among salaried workers, especially those at the higher income levels.[2] Schemes arranged by insurance companies were more responsible for the growth of inequalities than non-insured schemes. These different trends for manual and non-manual workers were only in part attributable to the more rapid increase in the employment of non-manual workers.

5. That in contrast to the position in 1961 (and apart from the civil service, the Armed Forces and the universities) tax-free lump sums were relatively uncommon in 1938 in respect of the commutation of either 25 per cent or from 25 per cent to 100 per cent of retirement benefits.[3]

If it were possible to translate these proportions into quantitative values and adjust the pre-tax income statistics since 1938 accordingly the results might well be remarkable. The statistics at the top of the income distribution would clearly be those most affected. To provide some substance to these general conclusions we now analyse briefly the trend in occupational pension schemes since 1938.

IV

Life Office Schemes[4]

According to the Ministry of Labour enquiry at the end of 1936, about 255,000 employees were covered for retirement benefits of

---

[1] For additional evidence on this point see *Minutes of Evidence, Royal Commission on the Civil Service*, Day 7, 1954.

[2] Of approximately 11,600,000 male employees in the private sector in 1956-7, about 86 per cent of salaried staffs and 20 per cent of wage earners had some private occupational pension cover. The difference was much more marked in schemes arranged by insurance companies than in non-insured schemes. This estimate was derived from an analysis of the Government Actuary's Report and other materials (see Titmuss, Richard M., *Government Pension Proposals*, Industrial Welfare Society, 1959, p. 6, and *Occupational Pension Schemes*, A Survey by the Government Actuary, 1958). See also figures given later for 1936-8.

[3] The Ministry of Labour survey of private pension schemes at the end of 1936 reported that commutation was exceptional and only in special circumstances approved by employers (*Ministry of Labour Gazette*, May 1938, Vol. XLVI, No. 5, p. 172).

[4] The following estimates, which are subject to a considerable margin of error, are derived from *Ministry of Labour Gazette*, May 1938; Bacon F. W. and others, 'The Growth of Pension Rights and their Impact on the National Economy',

some kind under insured schemes. Only 137,000 were non-manual workers and of this number it may be presumed that 21 per cent were women (assuming the same sex ratio as for insured and non-insured schemes combined). This means that about 108,000 men (described as administrative, clerical, sales, etc. staff) were covered at the end of 1936. From this figure a further deduction should be made for employees of transport, gas, electricity, hospital and education establishments transferred to the public sector of super-annuation after the war. It would probably be safe to say, therefore, that fewer than 100,000 male non-manual employees in the private sector (as defined for the post-war years) were covered by insured schemes before the war. Nearly all these employees were in contributory schemes; the enquiry reported that 248,000 out of the 255,000 employees were contributors.

For the post-war years the published information is even less detailed. It appears that by 1951 the total for manual and non-manual employees had risen to 1,400,000 (including 250,000 women). It is possible that a substantial proportion of the higher paid employees among the 1,150,000 men were (and still are) entitled to 100 per cent tax-free lump sum benefits. Much of this expansion of over 1,000,000 probably took place during the war; deferred pay and tax-free lump sums through insured schemes were arranged in place of salary increases thought to be inappropriate—as well as fiscally less rewarding—in wartime.

In 1956-7, according to the Government Actuary's survey, the total for insured schemes had risen to 2,666,000. By contrast with the position in 1936, the proportion of manual workers covered was extremely small. Thus, the great expansion during these twenty years in provision under insured schemes chiefly benefited non-manual workers.

According to the Life Offices' Association, the total number of employees covered at the end of 1960 was 3,483,000.[1] This represented an increase of nearly a million since the end of 1956 and was accepted by the Government as an indication of the success of such schemes in providing assured pension benefits for an ever-increasing number of employees.

The composition of these figures is, however, something of a

Journal of the Institute of Actuaries, Vol. 80, Pt. II, No. 355, 1954; _Occupational Pension Schemes_, A Survey by the Government Actuary, 1958; _Report of the Committee on the Economic and Financial Problems of the Provision for Old Age_, Cmd. 9333, 1954; and _Report by the Government Actuary on the National Health Service Superannuation Schemes 1948–55._

[1] United Kingdom only. In addition, 137,000 individual policies had been effected under the 1956 Finance Act provisions (_British Life Assurance Statistics 1956–60_, the Life Offices' Association and Associated Scottish Life Offices, October 1961).

mystery. In the first place, the life offices are unable to supply any classification by 'works' and 'staff', types of schemes, or manual and non-manual. Secondly, they cannot provide any information about age, sex and range of benefits. Thirdly, the premium figures cannot be separated into yearly premiums for group schemes, individual or Top Hat schemes, and back-service payments. Fourthly, it appears that the totals are made up of a mixture of different ways of defining and counting employees and other units. In general, it seems that withdrawing members who retain a paid-up pension (although the benefit may be very small) are still counted as 'employees'. Some offices do not, however, do so. Those that do—and they appear to be the majority—will contribute over the years to an accumulating amount of double-counting. Men who have had, say, four different employers by the age of fifty, and who have three paid-up policies in addition to being currently covered by a non-insured scheme, will appear three times in the statistics for insured schemes and once in the total for non-insured schemes.

Then again those who cease to be employees and become pensioners are still apparently counted 'in most cases' as employees. The benefits they receive are also counted in the current figures of 'pensions per annum in course of payment.'[1]

How many employees have lost their pension expectations (or have had them substantially reduced) in recent years as a result of change of employment, redundancy, take-over bids, strikes, and for other reasons? This question is likely to have affected manual workers to a considerably greater extent than salaried staff. It is an important factor in estimating what additions should be made to the personal income statistics for different income groups in respect of future pension benefits.

The published life office statistics show that for the years 1956–60 a total of £95,318,000 was refunded under pension schemes. It is not known how many individuals were involved. These withdrawals have been increasing rapidly. Between 1956 and 1960 the annual amounts refunded rose from nearly £11m. to over £25m. or by 130 per cent. The Secretary of the Association reports that so far as the number of employees leaving pension schemes is concerned, 'a rough indication is given by the difference between the total number of new employees covered less the increase for the year of the total number of employees covered'. Applying this method to the published statistics results in a total number of withdrawals for the four

---

[1] One leading office, it is said, does not however count pensioners as employees. This and other statements in these paragraphs about the statistics are taken from a letter to the author from the Secretary of The Life Offices' Association (dated December 18, 1961) in response to a series of questions.

years 1957–60 of 1,143,000. To this figure an unknown but substantial addition should be made for those transferring from one insured scheme to another, and for those who have been given paid-up policies and pensions but who are still counted as current employees. It is impossible to say what further corrections should be made for deaths in service, new pensioners, deaths among those with paid-up policies and pensions and other factors. Nor do we know what proportion of these withdrawals were subsequently covered for pension benefits under non-insured or public service schemes.

These gaps in the statistics and the possibilities for double- and even treble-counting are so considerable that it would seem wise to discount heavily the published statistics so far as they relate to the numbers covered. Until steps are taken by the life offices and the Government to arrange for more data to be made available we can only conclude (a) that the total numbers covered are seriously inflated, (b) that this inflation is more likely to have affected the lower income groups, (c) that, in consequence, a relatively higher proportion of the contributions and the increase in the life and pension funds should be credited as personal income to the higher income groups.

Private Non-insured Schemes

In 1936 there were 696,000 manual and 666,000 non-manual employees covered by private non-insured schemes (excluding those who received *ex gratia* payments). For comparability with post-war figures, it is probable that deductions should be made in respect of about 290,000 (non-manual) and 334,000 (manual) employees subsequently transferred to the public superannuation sector. This means that a total of about 738,000 employees (376,000 non-manual and 362,000 manual) of whom about 21 per cent were women were covered by private non-insured schemes at the end of 1936. Included in these figures are 131,000 non-manual and 50,000 manual employees of co-operative societies. About two-thirds of all non-manual employees covered by non-insured schemes were contributors and about one-half of all manual employees.

In 1951 the total under non-insured schemes was said to be 2,500,000 (including 400,000 women). For 1956–7 the Government Actuary's Report gave a figure of 2,333,000. No later information of a detailed kind has been published, and it is not possible to subject these crude data to the kind of examination given to the statistics for insured schemes.

It is apparent that, when subtractions are made for transport, gas, electricity, hospitals, nursing associations and co-operative societies, the proportion of employees (both non-manual and manual) in the private sector who had some retirement provision under insured and non-insured schemes in 1936 was very small.[1]

Public Schemes (including the civil service and the nationalized industries)

The Ministry of Labour 1936 inquiry did not include in its scope the schemes in operation at that time for certain classes of public employees. The main classes were: established civil servants, the Armed Forces, the police, firemen, teachers employed by public education authorities, and whole-time officers of local authorities. No comprehensive study has been made of the total number of employees in these classes who were covered in 1936 by some super-annuation or pension provision. It may be estimated, however, that, excluding the Armed Forces and the police, the total was in the region of 480,000.[2]

In 1958 the Government Actuary reported that nearly 4,000,000 persons employed in the public service and nationalized industries were covered in 1957. This figure excluded the Armed Forces. Although comparable figures are not completely available for the pre-war period, it is apparent that a very great increase took place in the number of employees with provision for pensions and super-annuation under public schemes. Some part of this increase was attributable to an expansion in the number of teachers and in the number of persons employed by the civil service and local government. But a very large part resulted from nationalization and the establishment of the National Health Service.[3] In 1936, for example, only 3,600 mineworkers and 7,100 administrative staff had any provision. At the beginning of 1960 the National Coal Board's schemes covered 586,000 mineworkers and 103,000 salaried staff. Large expansions also took place in transport and communications, electricity, gas and dock labour, while the creation of new public agencies such as the Atomic Energy Authority and air transport added further numbers. The National Health Service was providing for some 300,000 employees in 1955, a figure which represented a big increase compared with the number covered before the war. Manual workers, it seems, benefited more from nationalization than their opposite numbers employed by private firms who arranged their pension schemes with insurance companies.

It is clear that during this period of twenty years or so there took place a great expansion in the number of people provided with

[1] For insured and non-insured schemes combined there were fewer than 90,000 pensioners in 1936 under contractual arrangements.

[2] Political and Economic Planning, *British Social Services*, 1937, p. 123.

[3] We have shown earlier that if comprehensive statistics were available for public schemes in 1936 about 632,000 employees (a substantial proportion of whom were subsequently transferred to such schemes) would have to be deducted from the private sector to achieve comparability with the 1958 statistics.

some form of occupational pension. In round figures, the rise was from 2,000,000 to 9,000,000. The various estimates quoted do not, however, indicate the change in the annual amounts set aside by way of contributions and premiums by employers and employees (plus the investment income of the funds) nor the annual sacrifice of revenue due to tax reliefs for life assurance and superannuation contributions. As this expansion benefited the higher income groups to a much greater extent than manual workers, the amounts involved under both heads would no doubt show a more striking increase than the rise in the numbers covered if the data were available for comparative measurement.

The quantity and quality of information concerning contributions, premiums and other details which has been published by insurance companies, non-insured schemes and the Treasury is so appallingly meagre, however, that it is very difficult to do more than consider certain very general estimates.[1] These we must now examine as we have reached the point in this analysis of 'spreading income over retirement' when the effects on the income distribution statistics need to be explored. So far, we have shown the growing importance of this factor (or group of factors) since 1938. In what ways has this relatively new phenomenon of occupational pensions affected the statistics for different income groups, and what adjustments need to be made to the pre-tax income data for different ranges of income?

Mr Lydall is the only income statistician who has essayed the difficult task of making at least some adjustments to the statistics of statutory income. This he did as part of his general treatment of unallocated income which we have already referred to in the preceding chapter. We then left over for this chapter the particular problem of adjustments for pension schemes and life assurance.

'Allocated income,' said Mr Lydall, 'includes life assurance premiums paid by persons but it does not include life assurance premiums paid by employers on behalf of their employees. It includes pensions received, but it excludes almost all superannuation contributions—both by employers and employees—and the investment income of life and superannuation funds. All the excluded items are clearly of some benefit to the persons insured or covered by superannuation schemes; and although there may be some people who would not voluntarily save as much of their income as they are obliged to do under compulsory pension schemes, I think it is reasonable to treat the whole of this flow as forming part of personal

[1] This is part of the much wider problem of estimating personal savings and investment. Mr C. T. Saunders, in discussing this problem in 1954, drew attention to the inadequacies in the data and the need for more published information (Saunders C. T., *The Review of Economic Studies*, 1954-5, Vol. XXII, No. 58, p. 109).

income. At the same time, the value of current pensions from previous employers, which is already included in unallocated income, must be subtracted.'[1]

After briefly referring to the growth of schemes since 1938, Mr Lydall then gave estimates of employers' contributions for life assurance and superannuation, plus employees' contributions for superannuation, plus investment income of life and superannuation funds (after tax), minus pensions included in allocated income. He arrived at totals of £119m. for 1938 and £649m. for 1957 which were added to his table of adjustments to allocated income after tax.

It would seem that these estimates exclude notional contributions by the Central Government[2] and other public and private employers, notional contributions for pensions and other benefits for members of the Armed Forces, some element of back-service contributions under private schemes (paid by employers in the early years of schemes which were developed in the late 1940's and early 1950's), and payments under certain other more complex schemes referred to in Appendix D. Allowing also for sources of error in the statistics for insured and non-insured schemes, it would not perhaps be wildly wrong to add 15 per cent to the 1938 total for these excluded items and 25 per cent for 1957.[3] This would raise the totals to £137m. (1938) and at least £811m. (1957).[4] These adjustments mean the setting aside of £69 per employee in 1938 (2m. covered) and only £90 in 1957 (9m. covered). If allowance is made for the fall in the value of money and we only double the 1938 per capita figure this would result in a 1957 total of £1,233m.

If these figures are anywhere near the truth it would seem that the amount involved in 'spreading income over life' and converting income into capital through occupational pension schemes increased by about 900 per cent between 1938 and 1957. By contrast, the increase in the total of personal income before tax between 1938 and

---

[1] Lydall H. F., *op. cit.* (1959), p. 25.

[2] Between 1937–8 and 1960–1 the cost of superannuation benefits of retired civil servants rose from £9·8m. to £69m. The total staff employed in these years numbered 512,000 and 996,000 respectively. Emerging superannuation costs in 1970 and 1990, based on 1961 salary levels, are estimated at about £90m. and £125m. respectively (*Hansard*, H. of C., April 19, 1961, Vol. 638, Col. 111).

[3] Applying a notional contribution rate of 18 per cent to the civil service salary bill alone in 1938 and 1955 would account for £11m. of the 1938 addition of £18m. and £77m. of the 1957 addition of £162m.

[4] The *National Income and Expenditure Blue Book* for 1961 (table 24) shows that contributions by employers for 'life assurance, superannuation schemes, etc.', rose by 126 per cent between 1949 and 1957 (from £215m. in 1949 to £486m. in 1957 and to £569m. in 1959). 'Contributions by employees, individual premiums, etc'., rose less rapidly—from £323m. in 1949 to £559m. in 1957 and £654m. in 1959. Comparable figures for earlier years are not available.

L

1957 was 252 per cent.[1] Exchequer receipts from surtax rose from £70m. in 1938–9 to only £157m. in 1957–8 or 124 per cent.[2]

To correct for the deficiencies in the Board of Inland Revenue's income distribution tables it is necessary, therefore, to adjust the income before tax statistics by these amounts of £137m. for 1938 and £1,233m. for 1957. Further adjustments should in theory be made for the effects of taxation; for the additional benefits derived from exchanging reduced salaries and fees (or forgoing increments) for tax-free lump sums on retirement; for endowment assurances, education covenants, widows' and other benefits, the attributable parts of the investment income and appreciated assets of life assurance companies, and National Insurance contributions. An adjustment for the last item is a fairly simple matter and would be on the lines of that adopted by Mr Lydall.[3] This particular adjustment, taken alone, would to a fractional extent add to other factors making for a more unequal distribution of incomes in 1957.

Other detailed adjustments were made by Mr Lydall to the statistics for allocated income *after* tax—but not *before* tax—to take account of superannuation contributions and other forms of income and benefit. Accordingly, he fed back his estimates of £119m. (1938) and £649m. (1957) in respect of 'life assurance and superannuation' (as defined above). In doing so, he 'hazarded the guess' that 50 per cent of the employment income of the top 1 per cent income group was covered by a pension scheme in 1938; that 90 per cent was so covered in 1957; and that in each year the total superannuation contributions by employer and employee amounted to 15 per cent of the income covered.

These admittedly tentative assumptions may be criticized on at least five grounds:

1. The total estimates are too low, especially that for 1957 of £649m. for the reasons explained earlier.

2. The assumption of 15 per cent for *joint* contributions is much too low for 1957 for the top 1 per cent income group when account is taken of back-service payments by employers, widows' benefits, life assurance cover and a variety of other benefits referred to in this chapter and in Appendices D and E.[4] The total contribution to

---

[1] *NIBB*, 1960, table 22.

[2] *BIR, 83* and *103*.

[3] Lydall H. F., *op. cit.* (1959), p. 27. The higher income groups and the self-employed paid no contributions in 1938. Since 1948 they have done so, the self-employed (who constituted 40 per cent of the 350,000 surtax payers in 1958–9) paying higher contributions than employees. In 1959–60 tax forgone on National Insurance contributions by taxpayers amounted to £46m. Of this, 9 per cent was roughly attributable to the top 1 per cent of incomes—£2,000 and over (*Hansard*, H. of C., February 17, 1961, Vol. 634, Col. 189).

[4] After the discussion which followed the delivery of his paper to the Royal

superannuation benefits alone may amount to 25 per cent or more of total remuneration according to some authorities;[1] the Millard Tucker Report referred to employee contributions among higher paid staff of as much as 40 per cent of salaries.[2] The Report recommended a limit of 15 per cent for employees but none for employers except a warning that contributions should not be 'excessive'. A study of the report of the Ministry of Labour inquiry in 1936 shows that the average joint contribution was much less than 15 per cent.[3]

3. Account is not taken in the estimates of income covered for the top 1 per cent of the growth between 1938 and 1957 in tax-free lump sum benefits which, in individual cases, may have amounted to as much as £250,000.

4. If the analysis we have made of the Ministry of Labour inquiry in 1936 is broadly accepted then it would seem most improbable that as much as 50 per cent of the employment income of the top 1 per cent of incomes was covered by a pension scheme in 1938. The survey of *British Incomes and Savings* by Mr Lydall himself elicited proportions in 1952 for 'Managers, etc.' of 58 per cent; for income units with gross incomes of £1,500–£1,999 of 37 per cent; and for those with gross incomes of £2,000 and over of only 20 per cent.[4] All the evidence suggests that a significant development of 'top hat' and other individual superannuation, life cover and loss of office compensation schemes did not take place until the early 1940's.[5]

5. Mr Lydall further assumed that the share of the investment income of the life and superannuation funds accruing to the top 1 per cent group was 10 per cent in each year.[6] Although there is only scanty evidence to support this or any other assumption, the proportion could be much higher for 1957. The Board of Inland Revenue's income survey for 1949–50 provides some information

---

Statistical Society, Mr Lydall stated that 'in many cases the total pension contributions on behalf of higher paid employees' may be 'substantially in excess of 15 per cent of salary' (Lydall H. F., *op. cit.* (1959), p. 46).

[1] Innes E., *Journal of the Royal Statistical Society*, 1959, Vol. 122, Part 1, p. 43.

[2] Cmd. 9063, 1954, p. 41.

[3] *Ministry of Labour Gazette*, May 1938. Many studies of savings in the 1930's made no reference to the role of superannuation and pension schemes. The subject was not indexed in, for example, *Savings in Great Britain 1922–35* by E. A. Radice (1939). The absence of any serious study of the subject no doubt reflects the relatively small provision before the Second World War (apart from *ex gratia* payments about which practically nothing is known).

[4] Lydall H. F., *British Incomes and Savings*, 1955, p. 113.

[5] See, in particular, *Report of the Committee on the Taxation Treatment of Provisions for Retirement*, Cmd. 9063, 1954.

[6] The increase in 1938 in the funds of life assurance companies established in the UK was £38m. (average of £28m. for 1936–40). For life assurance and superannuation schemes combined, the net increase in funds rose from £238m. in 1949 to £708m. in 1959 (*NIBB*, 1961, table 24).

for the early post-war period.[1] For 26,700 persons allowed £100 and over for life assurance relief the average amount was £171. For 2,800 persons with net incomes of £10,000 and over the average amount of relief was £266. Just over 11 per cent of all the relief allowed exceeding £10 per person was taken by those with net incomes of £2,500 and over. The proportion was 14 per cent for those with incomes before tax of £2,000 and over—a group which, in 1949–50, broadly represented the top 1 per cent of incomes. These proportions would be higher if adjustments could be made to allow for (a) life assurance relief per family and not per person (b) life assurance for employees and their dependents paid by employers (c) the fact that the types of assurances effected by the higher income groups are often more profitable than those effected by the generality of taxpayers (d) the further fact that the proportion of investment income to premium rises with the age of the policy, i.e. proportionately more of the investment income would accrue to older people (e) developments since 1949–50, and (f) the rising rate of withdrawals under insured schemes in recent years which is likely to have affected in particular the lower income groups.

The whole subject of the ownership of pension funds, life funds and trust funds is extraordinarily complex, as Mr Lydall and Mr Tipping pointed out in their study of the distribution of personal wealth.[2] These funds now bulk large in the aggregate of personal net capital in Britain. From the review we have made of the growth of discretionary trusts, settlements and covenants of various kinds it seems that there must be a close interlocking in the ownership of these funds. In many respects, they are simultaneously involved in the variety of methods of splitting and spreading income-wealth and transforming income into capital.

If account could be taken of all these factors it might well emerge that by 1957 substantially more than 10 per cent of the investment income of the life and superannuation funds accrued to the top 1 per cent of income units, particularly if the units were defined in terms of families rather than persons.[3] Moreover, during the last fifteen years many types of associated schemes and trusts for 'spreading

---

[1] *BIR 95*. The extent to which the upper income classes make use of superannuation and life assurance was shown by the 1956 Oxford Survey. Annual premiums paid for life assurance ranged from £26 (gross income £1,000–£1,499) to £252 (gross income £10,000 and over). Superannuation contributions ranged from £26 to £167. These excluded contributions and premiums (notional or otherwise) paid by employers (Klein L. R. *et al.*, *op. cit.*, p. 313).

[2] Lydall H. F. and Tipping D. G., *op. cit.*

[3] In this connection it should be remembered that, according to Mr Lydall and Mr Tipping, the top 1 per cent of persons aged 20 and over owned in 1954 about 43 per cent of total personal net capital in Britain—excluding pension and trust funds (Lydall H. F. and Tipping D. G., *op. cit.*, p. 97).

income over life', 'splitting income between the kinship', and 'converting income into capital' have benefited from (as well as helping to cause) two dominant trends in the economy: first, the increasing share of total company dividends and capital gains from property development and other enterprises flowing into life, superannuation and trust funds and, secondly, the spectacular rise in the 1950's in the value of the assets of assurance companies and superannuation funds.[1] Between 1952 and 1960, insurance dividends in total increased by about 12 per cent a year, and investment income by 10 per cent a year.[2] For many reasons, these trends are likely to have had the effect of increasing the share of income-wealth accruing to the top 1 per cent of incomes.

These questions, like many others raised in this study, obviously call for more detailed and critical examination before all the necessary adjustments could be made to the official statistics. Further research, whether from the standpoint of social accounting, savings and investment, the distribution of income and wealth, or the restricted approach of this inquiry, would heavily depend on the availability of more basic data from the public and private sectors of the economy and, in particular, data that are more precisely defined than much of the material we have been handling. In recent years, the Government's figures of personal savings have been continually revised in a drastic and often bewildering fashion. Savings in 1952 were, for instance, put at £505m. in 1953, raised to £785m. by 1955, and then steadily reduced to £639m. in the 1960 edition of the National Income Blue Book. The cause of this uncertainty about the volume of saving is in part attributable to the inadequate data furnished by life assurance companies and superannuation funds. There are, of course, many other factors, which we cannot go into here, involved in the difference between total identified personal saving and total personal saving calculated by the Government statisticians as a residual of the income and expenditure accounts.

More relevant to the theme of this particular study, however, is the fact that at present we are quite remarkably ignorant of what is happening to the distribution of income-wealth in relation to the life span of individuals and the generation span of family and kinship; nor have we yet recognized what effects these developments are having or may have on the institution of progressive taxation.

The Phillips Committee in its 1954 Report, explaining that 'a substantial proportion of the cost of occupational pension schemes is borne by the State', also referred to an estimate that 'the annual

---

[1] According to an estimate quoted by Mr W. Nursaw, £100 invested in 1951 in the Legal and General Assurance Company would have been worth £1,744 to a shareholder in 1961 (The *Observer*, September 17, 1961).

[2] Nursaw W., The *Observer*, December 10, 1961.

sacrifice of revenue' involved in tax reliefs for superannuation schemes alone 'may be of the order of £100m.'[1] From the context of the work of this Committee and the Millard Tucker Committee it would seem that this estimate was made in the early 1950's. It was not explained what it covered and how it was made. Clearly, it could only have been a guess, for the Board of Inland Revenue had no knowledge at the time of (a) the extent of non-taxable *ex gratia* lump sums annually being paid (b) whether tax-free lump sum benefits already contracted for and maturing in the future involved a total 'spread of income' of £1,000m. or £2,000m. or any such figure[2] (c) whether contributions, as a business expense, by employers for their higher paid staff in respect of retirement benefits, life cover[3] and widows' and dependents' provisions were of the order of 20 per cent of salaries, 40 per cent or more and (d) to what extent tax forgone in respect of the whole field of retirement provisions (including the investment income of superannuation funds) was linked to the problem of tax and estate duty forgone on discretionary trusts, covenants, family settlements and other schemes for splitting income and converting income into capital. How little is now known about these matters is apparent from the questionnaire we submitted to the Board of Inland Revenue (see Appendix A).

Two years after the Phillips Committee had reported, the Finance Act of 1956 made provision for retirement annuities for the self-employed and controlling directors. The tax reliefs allowed were estimated by the Treasury to cost eventually an additional £50m. a year.[4] This development, together with an even faster rate of growth during the 1950's in 'top-hat' and insured schemes, discretionary trusts and similar arrangements for a relatively small section of the population, may conceivably have doubled or trebled the estimate of the 'annual sacrifice of revenue' made early in the decade. The opportunities for reducing taxation were many and various for, as Mr Nortcliffe pointed out in a survey of personal income in nine countries, 'the United Kingdom is probably unequalled in the

---

[1] Cmd. 9333, 1954, pp. 62 and 64.

[2] Supposing 100,000 or 200,000 individuals with 'rights' to tax-free lump sums of £10,000 each.

[3] According to the Government Actuary's Survey in 1957, employers paid additional contributions for life assurance, for instance, in respect of 38 per cent of the members of insured schemes (*Occupational Pension Schemes*, 1958, p. 19). Mr Copeman's survey of executives in 1957 showed that 96 per cent of those interviewed were covered by their employers for life assurance (Copeman G., *Promotion and Pay for Executives*, 1957, p. 198).

[4] *Hansard*, H. of C., July 10, 1956, Vol. 556, Col. 292. By the end of 1960, about 137,000 individual policies were in force, and tax forgone could only have amounted to about one-quarter of the Treasury estimate (Life Offices' Association, *British Life Assurance Statistics 1956–60*).

thoroughness of its reliefs for contributions towards retirement and survivor benefits'.[1]

Life assurance relief, originally regarded as a privileged form of saving singled out for preferential treatment—a 'special indulgence' —for the lower middle classes because they had no pensions and found it hard to save for their old age, was costing the taxpayer £9½m. in 1938–9.[2] By 1959–60 this 'special indulgence' (no longer restricted to the lower middle classes) was costing £49m. (excluding relief on premiums paid by employers); about one-seventh of it was being received by the top 1 per cent of incomes.[3] By then, as we have already pointed out, life assurance had become inextricably mixed up with the development of discretionary trusts, education covenants, 'top hat' retirement benefits, the 'purchase' of tax-free lump sums, separate pensions for widows and other dependents, and a variety of other schemes for estate duty saving. Some of the ways in which life assurance has been deployed in recent years to reduce taxation and estate duty are illustrated by Mr Lynes in Appendices D and E. The Board of Inland Revenue is unable to estimate the amount of tax and estate duty forgone on life assurance premiums paid for by employers for employees and their wives and children.

It is indisputable that many of these arrangements—whatever their technical label and whoever actually pays the contribution, premium or charge—are not simply and solely irrevocable saving for old age. Yet we stoutly maintain the myth that the fiscal system does not and should not encourage savings or deferred consumption apart from the one exception of specific and irrevocable provision for old age. In July 1958 the Chancellor of the Exchequer said: '. . . the principle on which we work—and I think it is a sound one— is that Income Tax should fall impartially on all income from all sources, regardless of whether that income is spent or saved.'[4] To do otherwise would be to subsidize a privileged group of taxpayers by throwing a 'burden' on the rest of the population. In other words, redistribution would take place in favour of those taxpayers who were in a position to 'arrange' to defer consumption. This, as we have seen, is what has actually been happening on an increasing scale since 1938, partly because of the difficulty of distinguishing between saving in general and irrevocable saving for 'old age'. Much

[1] The other countries were: Australia, South Africa, Canada, USA, France, Netherlands, Western Germany and Sweden (Nortcliffe E. B., *British Tax Review*, September 1957).

[2] Cmd. 9063, 1954, and *Hansard*, H. of C., February 17, 1961, Vol. 634, Col. 189.

[3] *Hansard*, H. of C., February 17, 1961, Vol. 634, Col. 189.

[4] *Hansard*, H. of C., July 2, 1958, Vol. 590, Col. 1501. See also the discussion by the Financial Secretary to the Treasury on fiscal benefits and social service benefits (Cols. 1485–7) and inequitable tax burdens (Cols. 1534–5).

of it has been operating on what we can only describe as the 'most-favoured-taxpaying-family' principle.

The purpose of this excursion into a complex area of economic behaviour was a limited one. No attempt has therefore been made to enlarge on some of the fundamental issues thrown up by this cursory survey of changes in the time and kinship pattern of consumption and saving. They are important in many respects. Among those which are the concern of this study two may be singled out. The first is that, considered as a whole, the developments we have surveyed constitute one of the major factors in the erosion of the tax base since 1938. In consequence, the conventional frame in which the nation's income statistics are cast is increasingly presenting a delusive picture of the economic and social structure of society. Secondly, these developments, in conjunction with the growth in the resources and power of the financial institutions through which they are operating, are making for greater inequalities in the distribution of incomes from the perspective of the life span of the individual and the family

CHAPTER 8

Benefits in Kind

'WE feel bound to say in conclusion,' said the Royal Commission on Taxation, 'that the provision of untaxable benefits in kind is capable of becoming an abuse of the tax system. The reason is not simply that high taxation gives a special attraction to all forms of untaxed receipt. Modern improvements in the conditions of employment and the recognition by employers of a wide range of obligation towards the health, comfort and amenities of their staff may well lead to a greater proportion of an employee's true remuneration being expressed in a form that is neither money nor convertible into money.'[1]

In common with other fiscal issues discussed in this study, the problem of untaxed benefits in kind is not peculiar to the United Kingdom. During the 1950's, many detailed and wide ranging surveys were carried out in the United States, France, Japan and other countries. They all showed that the growth in the provision of benefits in kind is not solely a consequence of high taxation. Like the development of pensions through insurance companies and pension funds, they cannot be regarded as an isolated phenomenon. They have to be seen in their social context and as part of a general shift from contract to status. In the words of Father Harbrecht's conclusion: '. . . a man's relationship to things—material wealth— no longer determines his place in society but his place in society now determines his relationship to things.'[2] So far as employees of the corporation are concerned, this trend towards a paraproprietal society has been suggested by a number of recent American studies. On one element they are agreed: that the phenomenon of benefits in kind was relatively uncommon thirty or forty years ago. Such books as William H. Whyte's *The Organization Man* (1956), Alan Harrington's *Life in the 'Crystal Palace'* (1960) and Paul Harbrecht's *Pension Funds and Economic Power* (1959) illustrate the extent to which the 'labyrinth of benevolence' now affects the liberties as well as the

[1] *Final Report*, Cmd. 9474, 1955, pp. 71–2.
[2] Harbrecht P. P., *Pension Funds and Economic Power*, 1959, p. 287.

standard of living of many of those who now work in the private
company state in America.

The tax treatment of corporation benevolence has also attracted
the attention of many American economists, statisticians and
lawyers. Mr Macaulay, for example, in studying the growth of
fringe benefits (a broader concept including cash benefits) quoted
figures to show that all fringe benefits accounted for about 15 per
cent of payroll costs in 1949 and 24 per cent in 1957.[1] In an attempt
to classify these benefits, he provided a list of about sixty distinctive
ones ranging from business yachts to golf instruction. In discussing
the effects on patterns of consumption he referred to an estimate
that 'in large cities like New York, Chicago and Washington, at any
moment over one-half the people in the best hotels, night clubs, and
restaurants are paying for the services they receive *via* the expense
account'.[2]

Considered individually, the value of many benefits in kind may
seem to be small and trivial. In the aggregate, however, and from
the viewpoint of public finance, they may be quite considerable. This
appears to be the case in America, judging by the formidable indict-
ment 'Study on Entertainment Expenses' drawn up by the Treasury
in 1961 and based on 38,000 or so tax returns.[3] Professor Ratchford,
observing that the trend towards the use of fringe benefits as a form
of compensation is the reverse of what which took place after the
feudal period, and assessing the effects of this trend on American
tax revenues, was led to comment: 'Perhaps the time will come when
the individual unfortunate enough to receive all his wages in money
will have an impossible tax burden.'[4]

To cite these studies from America (where taxation is *thought* by
the British to be much lower) does not mean that their conclusions
necessarily apply to the United Kingdom. They have been quoted
to illustrate a general trend in modern productive systems, and
because the subject has received practically no serious attention in
this country. Apart from the few pages devoted to the subject in the
Royal Commission's Report and occasional journalistic essays about
expense account executives, the phenomenon of benefits in kind has

---

[1] Macaulay H. H., *Fringe Benefits and Their Federal Tax Treatment*, 1959,
pp. 8–11. According to an estimate by the Division of Program Research, Social
Security Administration, total expenditures in the USA for fringe benefits as
defined in *Research and Statistics Note No. 2*, 1959, amounted to 25·6 per cent of
payroll in 1957 or $1,151 per man-year.

[2] *Op. cit.*, p. 58. Problems of equity arise from different national concepts and
tax definitions of fringe benefits. American business visitors to Japan find, for
example, that the services of geisha girls are supplied by Japanese employers as
a fringe benefit; these are not considered in Japanese fiscal law as income.

[3] For an English summary see *The Economist*, June 3, 1961, p. 1002.

[4] Ratchford B. U., *Journal of Finance*, VII, 1952, p. 211.

mostly been ignored in any discussion of incentives to earn, incentives to change jobs, and economic behaviour in general. This neglect is particularly noticeable among those who have studied the changing distribution of incomes since 1938. Most of them do not even mention the subject.

The reason no doubt is the lack of any comprehensive information not only about the receipt of such benefits by the top income groups but by all income classes. We have been similarly handicapped. This study could not, however, be brought to an end without some reference to the question. We have, therefore, been forced to draw together a limited amount of scattered evidence most of which, it so happens, affects the statistics for the higher income groups. It should not be inferred from this, however, that benefits in kind are not received by large sections of the lower income classes. This may well be the case, although what is covered by the term 'benefits in kind' may easily shade into a multitude of minor acts of avoidance and evasion—perquisites in the form of free or cheap goods from factories and shops, free travel for various classes of employees, tools, clothing, free coal for miners and so forth.

This point was made by Mr G. Paine of the Board of Inland Revenue in the discussion on Mr Lydall's paper in 1958.[1] Regrettably, however, there is little information available which would allow even approximate estimates to be made of the potential effects on the statistics of pre-tax incomes for the lower groups—apart from the adjustments made by the Central Statistical Office for domestic servants and agricultural workers. The figures for individual income units are unlikely, it may be supposed, to be affected to any substantial extent. In the aggregate, however, the addition of such benefits could be significant. For the higher income groups, both figures may be important; those for individual units as well as for groups of units.

We are not, it must be emphasized, directly concerned here with the question whether and to what extent these benefits for all income groups should be taxed. Our primary interest is in the effects of the growth of benefits in kind—not as an isolated phenomenon but in conjunction with the other social and economic forces we have surveyed—on the pattern of income distribution since 1938.

From an assessment of such evidence as is available for the more important (in terms of money values) benefits in kind we reached two conclusions. These may be stated here. First, that in the United Kingdom as in other modern economies there has been a marked growth in the provision of benefits in kind during the last twenty years or so, and, secondly, that for many groups in the occupied population the receipt of such benefits rises sharply with income.

[1] Paine G., *Journal of the Royal Statistical Society*, 1959, Vol. 122, Pt. 1, p. 39.

We are, therefore, led to ask: to what extent do the Board of Inland Revenue's statistics of income before tax—income brought under review—include the value of these and other benefits in kind? The general impression we have is that most benefits are not included in full; where, in individual cases, some element is included it is unlikely to represent the true costs to the employee; that is, what it would cost him in current market values to provide the benefit himself.[1] Benefits in kind are not as a general rule taxed on current market values but on a different basis.[2] The principle has been laid down in a number of cases before the Courts that a person is chargeable not on what saves his pocket but what goes into his pocket. As regards benefits in kind, what goes into a person's pocket is the value of the perquisite received in his hands, and this, of course, is not necessarily the same as the cost of the perquisite.[3]

The Royal Commission came to the conclusion in 1955 that 'Benefits in kind are very widespread and of the most diverse nature.'[4]

The Report went on to discuss the legal and administrative problems involved in tax treatment:

[1] In the Board's Memorandum No. 8 'Benefits in Kind' submitted to the Royal Commission on Taxation and dated May 1951 it was stated:

'Benefits in kind are defined for the purposes of this legislation (Part IV, Finance Act, 1948) as including the provision of living and other accommodation, entertainment, domestic or other services or other benefits or facilities of whatever nature. The measure of liability is normally the cost to the employer of providing the benefits in kind. If the benefit is the use of an asset which remains the property of the employer, however, the initial cost of acquiring or producing the asset is not taken into account; liability falls on the annual value of its use or on the rent or hire paid for it by the employer, whichever is the greater, in addition to any current expenditure on the asset, such as petrol for a car. If the benefit is the provision of living accommodation the amount of the liability is the rent paid for the premises by the employer or the net Schedule A assessment, whichever is the higher, in addition to the cost of any services such as domestic staff, gardeners, or any outgoings on the premises. By concession the full charge may be abated if the house is patently too large and old-fashioned.'

The Memorandum went on to specify certain benefits in kind which are exempt from charge by statute.

[2] The Chairman: 'Now, if you get a benefit in kind under the 1948 Act, you are not taxed on the market value of it but on what it costs your employer to give it to you; is that right?'

Sir Alfred Road on behalf of the Board: 'Yes.'

The Chairman: 'Which may be very much less than the market value?'

Sir Alfred Road: 'Yes.'

(*Minutes of Evidence, Royal Commission*, 5920–1, July 8, 1954.)

[3] Illustrated in the case of a managing director granted the privilege of applying for unissued shares at par and, also, in the more famous case of the employer who gave suits value £14 5s each to twenty-two employees. The value on which tax was chargeable was agreed at £5 being the value of the suit second-hand as soon as it was received (see *British Tax Review*, March-April 1960, p. 130).

[4] Cmd. 9474, 1955, p. 68.

A benefit in kind may be convertible or inconvertible. It is convertible if it is of such a nature that the recipient can turn it into money: for instance, employees of many factories which produce consumer goods are allowed a free or reduced rate issue of such goods; the coal miner was, traditionally, allowed a free issue of coal. A benefit is inconvertible if it is of such a nature that it can only be enjoyed by the recipient: for instance, the domestic servant's board and lodging, an employee's free meals, a railway-man's free travel, a bank manager's living accommodation at the bank, a teacher's free education for his children. Subject to certain special legislation in 1948, the inconvertible benefit is not treated as taxable income apparently because it is not regarded as being equivalent to money's worth. At any rate there is no ready means of assigning to it a monetary equivalent; and, whatever administrative expedients might be adopted, the task of assessing the annual value of every such benefit and collecting tax upon it strikes us as a singularly formidable and laborious one. This is the fundamental difficulty in any treatment of the matter that seeks to be consistent. And the fact that non-convertible benefits are not taxed has led, not unnaturally, to a relaxation of the rule about convertible benefits. Where the volume of goods allowed to an employee by way of perquisite does not exceed what is reasonable for his own consumption, no tax charge is made in respect of them. Nor is tax charged on cash payments which coal miners receive in lieu of their former perquisite of free coal. The idea seems to be that perquisites of this kind are primarily intended for the con-sumption of the recipient and his family and, though they are of a nature that is convertible, the general understanding is that they are for personal enjoyment. They are therefore analogous to the inconvertible benefit. Since, however, the concession depends upon the volume of goods not being larger than would be needed for personal requirements, it does not go so far as to make it possible for any one individual to escape anything beyond a very small amount of tax.

On the other hand a taxable benefit arises whenever an employer discharges a pecuniary liability of an employee that he has not incurred in the course of performing the duties of his employment. Usual instances are the payment of the rent of a dwelling-house where its occupation is not part of the employment, or the provi-sion of a season ticket for travel from home to work. Sometimes children's school fees or holiday expenses are provided. This rule cuts across the distinction between convertible and inconvertible benefits, since the payment of a particular bill for a man gives him no option between personal enjoyment and realization. But any different rule would be contrary to common sense; and, pre-sumably, the same difficulties of valuation do not occur as those that tend to defeat the assessment of the ordinary inconvertible benefit. However, liability is not attached to an employee's occupa-tion of a house if the occupation is incidental to the duties of his employment, as it would be in the case of a lodge-keeper or level-crossing keeper or a colliery manager required to live in the neighbourhood of the mine.

What we have said is sufficient to illustrate the very great difficulty of applying any logically consistent treatment to this subject. Theoretically, all benefits in kind received in the course of employment and attributable to it are a form of remuneration and should rank as taxable income, since otherwise one taxpayer's income is not equitably balanced against another's. No doubt an inconvertible benefit has not got a monetary value in exchange, but that does not seem to us to be the same thing as saying that it does not possess any value at all which should be taken into account in a computation of income.[1]

The special legislation to which the Commission referred was the Finance Act of 1948 which made benefits in kind chargeable to tax as part of the remuneration of directors of trading companies and employees whose gross remuneration (salary plus expenses payments and benefits in kind) amounted to £2,000 per annum. A connected problem, which the Commission also discussed, was the legislation affecting personal expenses. Under the Finance Act of 1948, it was laid down that for directors and employees affected all sums 'paid in respect of expenses' should be treated as part of the assessable income.[2] The recipient was then free to make good such deductions from that income as he could show to be represented by money expended wholly, exclusively and necessarily in performing the duties of the office or employment.

A great deal of case law has developed in respect of Schedule E expenses as well as of the definition of inconvertible benefits in kind.[3] In general, the trend appears to have been, both in law and administrative practice, to admitting more personal expenses as deductible and more benefits as inconvertible. In August 1961, the Chancellor of the Exchequer, concerned about many allegations of abuses in the system, attempted to clarify the practices and rules built up by the Board of Inland Revenue to deal with the legislation of 1948. These were set out in Notes issued by the Board.[4] Apart from one or two

---

[1] Cmd. 9474, 1955, pp. 67–8.

[2] They had been chargeable before, but until 1948 the onus was upon the Revenue to show that a sum received by a director or employee had been spent on something for which he was not entitled to claim an allowance, 'and that onus was one very difficult to discharge' (Sir Alfred Road and Mr E. R. Brookes, on behalf of the Board, examined by the Royal Commission. *Minutes of Evidence*, July 8, 1954). These Minutes also explain some of the reasons for the 1948 legislation and the fact that after 1948 the Board did not know the amounts of money involved because of the system of local dispensations.

[3] See, for example, Monroe H. H., *British Tax Review*, March 1959.

[4] Board of Inland Revenue, *Notes on Expenses Payments and Benefits for Directors and Certain Employees*, No. 480, 1961.

minor changes, they restated the administration of the law, and were welcomed by the Institute of Directors as indicating a more liberal attitude towards extra allowances and expenses.[1]

However that may be, we come back to the primary concern of this study that it is impossible to tell from the Board's statistics since 1938 what benefits in kind have or have not been included in pre-tax income, what values have been placed on such benefits, and how the different income classes have been affected. No detailed and comprehensive information has been published on these questions. We thus do not know how the legislation of 1948 has worked in practice.

If it could be assumed that benefits in kind (including the value of expense allowances) were distributed more or less equally amongst the whole tax population then these questions would not merit further inquiry. But we cannot make this assumption. All the published evidence points the other way. Indeed, it is conceivable that the distribution in money values of all categories of benefits in kind may now be more unequal than post-tax income defined in conventional terms. If this be the case, then account would have to be taken of the further facts that, first, the value of these benefits to the individual income unit was substantially greater in the 1950's than in 1938 and, second, that such increments were not fully reflected in the statistics of pre-tax income.

The impression gained from reading all the income distribution studies since 1938 is that the authors, while accepting the fact that the official statistics largely exclude benefits in kind (other than those of domestic servants and agricultural workers), conclude that such additions to income are unimportant from the viewpoint of their analyses. We consider, on the other hand, that for certain classes and income groups they may be of substantial importance. We summarize below various scattered pieces of evidence for a number of the more significant categories of benefits in kind.

Expenses

Deductible expenses under Schedule E rose from £128m. for 1954–5 to £157m. for 1958–9. According to the Board no information is available for any earlier year,[2] or for expenses under Schedule D. Nor is it known how these deductions were distributed among different ranges of income.

---

[1] *The Times*, August 5, 1961. Following strong protests by the Institute of Directors in 1954 about 'inquisitions' on expenses, the Board, it was said, adopted a more liberal interpretation of the legislation (*The Times*, June 17, 1954 and Copeman G., *op. cit.*, pp. 200–4).

[2] *BIR 103* and *Hansard*, H. of C., May 20, 1960, Vol. 623, Col. 152.

Housing

In a sample of sixty-seven companies, described as reasonably representative, investigated by the British Institute of Management in 1960, all but ten helped their employees with house purchase, either by some kind of loan or by acting as a guarantor.[1] This usually involved the company in a guarantee for about 20 per cent of the purchase price. A minority of the companies only helped their senior staff to buy a house and some only the administrative staff, as distinct from the hourly paid factory workers. A number of companies had set up housing associations, some of them quite large (e.g. the English Electric Company's association owned 1,536 houses and BOAC 1,279). More than half the companies owned property which they let to employees, the rents being in some cases merely nominal. A substantial number of companies helped with removal expenses, and some companies met the cost incurred by an employee in selling his old house and buying a new one. Of the sixty-seven companies, fifty-two reported that they would, in most instances, pay some of the cost involved in actually selling one house and buying another.[2]

The Economist, discussing in 1961 the sale of expensive houses in London and the role of personal overdrafts guaranteed by employers, said, 'This is a perquisite that a growing number of large companies are offering their executives; other companies go the whole hog and lend employees the money themselves.'[3] A few years earlier it had commented on 'the multiplicity of service tenancies that exist today'.[4]

A memorandum addressed by the tenants' association of Dolphin Square, London, the largest block of flats in Europe, to the Minister of Housing and Local Government in 1960 complained: 'Property companies, speculators and landlords were aware of the shortage and were aware that the number of businessmen and others whose rents were paid as an "expense" was growing.'[5] Similar allegations were made in the House of Lords by Viscount Monsell of flats being

[1] Summarized in *The Times*, February 22, 1961. See also the results of a questionnaire on 'additional benefits' for staff (including directors) earning over £2,000 a year sent to twenty private firms and four nationalized bodies by the Royal Commission on Doctors' and Dentists' Remuneration. Ten employers gave loans; two provided rent-free accommodation; eleven others had a variety of arrangements for providing accommodation in special circumstances—varying from artificially low rents to economic rents (Cmd. 939, 1960, App. G).

[2] In some cases of housing assistance it has been held that such 'emoluments' are not taxable under Schedule E. See, for example, the House of Lords judgment in the case of an employee of ICI Ltd, reported in the *British Tax Review*, January-February 1960, p. 55.

[3] *The Economist*, March 4, 1961, p. 874.

[4] *The Economist*, February 16, 1957, p. 576.

[5] *The Times*, September 20, 1960. This property was involved in the *Affairs of H. Jasper & Co Ltd*, Report by Mr N. Faulks to the Board of Trade, 1961.

let on expense accounts at £6,000 a year, and by Mr Wyatt in the House of Commons of flat premiums of £20,000 and £30,000 being paid by employers.[1] A new five-storey block of flats 'designed specifically to provide chairmen of companies with attractive apartments in the centre of the City of London' was described and depicted by an artist on the city page of *The Times* on April 16, 1960.[2] It was said to be one of the current development projects of an investment trust.

It may be that some of these additions to income are chargeable to tax at current market prices. On the other hand, houses and flats provided by employers may be let at 'concessional rents'.[3] Whatever the reasons for an increasing amount of property being let in the name of companies, it does seem doubtful whether these developments are fully reflected in the Board of Inland Revenue's pre-tax income statistics. Where the terms of an employee's contract are that he receives a cash salary plus free board and lodging he is liable to tax on the cash salary only.[4] This follows from the general principle of the law that a benefit in kind received by an employee from his employer is not assessable to tax if it can only be enjoyed by the recipient and is not capable of being converted by him into cash.

What is certain is that the value of the housing 'subsidies' enjoyed by various categories of council tenants is not reflected in the statistics of pre-tax income. In the aggregate, this could be of importance to the bottom income groups. But in terms of the individual income unit the absolute value of such benefits is likely to be small in relation to housing benefits in kind received by individual units in the higher income classes.

Cars and Travel

All writers on the subject of fringe benefits in America as well as in the United Kingdom are agreed that the value of 'business' cars to those who drive them is greater in practice than the taxable benefit. Account has to be taken particularly of the costs met by employers for repairs, insurance, maintenance and running.

[1] *The Times*, June 1, 1960 and March 1, 1961. Advertisements in recent years in *The Times* and other papers by employers and appointments consultants sometimes specify, as part of the emoluments, 'house, car, non-contributory pension scheme, etc.' See, for example, issues of *The Times* for August 18, 1959 and January 15, 1960.

[2] See also *The Times*, October 12, 1961, advertising leases at rents from £1,250 to £1,500 per annum.

[3] This was the term used by Mr E. B. Nortcliffe in the Unilever Journal *Progress* in discussing the growth of fringe benefits (Spring 1959, p. 76). See also Copeman G., *op. cit.*, p. 206.

[4] Letter from the Secretaries' Office (Taxes), Inland Revenue, February 13, 1959.

M

In 1955 it was assumed by the Government statisticians that just under half of all new cars were bought by business firms, and that about one-fifth of the passenger services of British Railways were provided for businessmen and civil servants.[1] At the end of 1956 during the period of petrol rationing the Minister of Fuel reported that four out of five cars on the roads were owned by firms.[2] Lord Kindersley, chairman of Rolls Royce, said that 'most of their cars were bought by companies'.[3] Among employees and directors in the top 1 per cent of incomes it appears to be a general practice to provide either cars for their own use or chauffeur-driven cars or car services from a company pool or rental system.[4]

In 1961 the Government limited the capital allowance on cars bought as capital assets by business firms and professional people to a maximum of £2,000 for each car. So far as its use was concerned, the amount on which tax was payable was the proportion attributable to private use of the annual value of the use of the car fixed at 12½ per cent of the original cost (which assumed a life of eight years) and the running costs.[5]

Education
No discussion of the changing distribution of income and wealth can today leave out of account the value of special educational benefits, direct and indirect and in cash and in kind, received by a proportion of the population. The subject, in its broader and more fundamental aspects, lies outside the scope of this study. Nevertheless, we have had to refer more than once in earlier chapters to some of the ways in which expenditures on private education are closely related to the manipulation and rearrangement of income-wealth, and how such processes affect the conventional statistics of income. As we have shown, these expenditures or benefits, direct and indirect, take a variety of forms: direct payments or grants to educational establishments; the payment of premiums on endowment policies for educational purposes; the use of covenants and discretionary trusts; concessionary loans; charitable gifts to selected schools; rate relief, education trusts, scholarships and so on.

Some of these forms or methods are used by employers to channel educational benefits in favour of the children of selected employees. In recent years, they have been increasingly used by the larger cor-

[1] *The Economist*, June 9, 1956, p. 1011.
[2] *News Chronicle*, November 30, 1956.
[3] *The Times*, May 5, 1961.
[4] See Copeman G., *op. cit.*, p. 204; *Report of the Royal Commission on Doctors' and Dentists' Remuneration*, Cmd. 939, 1960, App. G on Additional Benefits in Industry and Commerce; and *The Times*, August 28, 1956.
[5] Finance Act 1961 and Board of Inland Revenue, *Notes on Expenses Payments and Benefits for Directors and Certain Employees*, No. 480, 1961, p. 10.

porations and although in part designed at the secondary school and higher education levels for recruitment purposes, they represent from the angle of income distribution a benefit in kind. The receipt of such benefits is not reflected at all or only to a fractional degree in the statistics of pre-tax income.

It is impossible to say how widespread these various arrangements are today, and to what extent some income groups benefit more than others.[1] Moreover, from the viewpoint of the income distribution statistics the whole problem is complicated by the confusion that surrounds the treatment of the income of minors and what, at different levels of income, constitutes an 'income unit'. Even to suppose that only 10 per cent of children at public schools benefit, directly or indirectly, from arrangements of one kind or another could affect the figures for the top 1 per cent of incomes if the value of these benefits is not fully represented as income of the parents in the Board's tables.

Two points can, however, be made with assurance, though they cannot be supported with precise statistics. First, these developments have taken place in the last ten years or so and are, in many respects, connected with the problem of recruiting staff at the higher managerial, technical and scientific levels. Others are designed to benefit selected employees and selected children.[2] All the evidence suggests that such 'benefits in kind' were far less common in 1938. Secondly, it can be said that these benefits are now highly selective and, from the point of view of the future careers of the children, highly important. The number and proportion of employees, salaried and hourly paid, who derive some benefit from these developments no doubt remains relatively insignificant. Even so, if they are not representative of all employees, they could now be substantial enough to influence the income pattern of a small group of income units.

Meals and Entertainment
Subsidized or free meals at work were by 1960 being provided for

---

[1] Whether grants were given for the education of employees' children was one of the questions asked in the survey of 'additional benefits in industry and commerce' in 1958 undertaken by the Royal Commission on Doctors' and Dentists' Remuneration. The enquiry related to all staff (including whole-time directors) earning over £2,000 a year, and it was addressed to twenty-four large employers. Of these, five (or 22 per cent) made some provision: two employers had educational trusts, two made loans, and one had a trust providing competitive scholarships (Cmd. 939, 1960, App. G).

[2] The Rolls Royce Educational Trust is an example of a number of such trusts set up in recent years. It was established in 1956 to help pay the school fees of salaried employees with two or more children. Wage-earners were excluded (*News Chronicle*, April 25, 1956).

large sections of the employed population.[1] In 1938 the position was very different. The extension of this benefit in kind for those who had their mid-day meals at work was greatly stimulated by the growth of the luncheon voucher system. This development chiefly benefited non-manual workers. By 1958 some 750,000 coupons were being issued each week and during 1956–8 the turnover doubled to a total of about £5m. a year.[2] By then they were becoming, as *The Economist* remarked, 'almost negotiable instruments'.[3] In 1959 the Government limited this non-taxable benefit to 3s a day and ordained that the vouchers should be non-transferable and used for meals only.[4] This led to some demand that those who had neither a staff canteen nor luncheon vouchers should be allowed a 15s a week tax-free allowance.

Associated with the growth of this category of benefits in kind during the 1950's was the purchase of spirits and imported wines on business account. In 1955 the Government raised this item from 5 to 10 per cent of estimated consumers' expenditure on wines and spirits—from £16½m. to £33m. a year.[5]

Little can be said on any factual basis about what the Lord Mayor of London described in 1960 as 'good living on the expense account'. 'The worst offenders,' he added, 'were sometimes the biggest organizations and it was often the small man who had to suffer in consequence.'[6] A survey of the economics of London's night clubs and late night restaurants in 1959 gave some substance to the Lord Mayor's views. Most of the bills for these forms of entertainment were paid by firms and not by individuals. 'Only Exchequer *largesse*, paid directly via the tax-free expense account, keeps the night club industry ticking at all.'[7]

Other forms of entertainment also raise questions about the legitimacy of charges against a company's profits. Sir Egbert Cadbury, in a letter to *The Times* in 1962, drew attention to the 'tendency of large industrial undertakings to buy up the best stretches of

---

[1] A survey of fifty-five firms in 1958 showed the employers' costs of canteens at £5 14s 0d per employee per year (W. Durham, *The £ S D of Welfare in Industry*, Industrial Welfare Society, 1958, p. 21).

[2] Report of the National Food Survey Committee, *The Times*, April 17, 1958.

[3] *The Economist*, February 16, 1957, p. 576.

[4] *BIR 102*, App. IX, list of extra-statutory concessions.

[5] *NIBB 1956*, table 22 and p. 68.

[6] *The Times*, November 30, 1960. See also statement in *Principles in Practice*, published by the Bow Group in 1961, concerning 'lavish non-taxable lunches' for executives (p. 23).

[7] *The Economist*, April 11, 1959, p. 105. According to Mr P. Raymond, at the time one of the leaders of the strip-tease club and theatre industry, 'many firms enrol their executives so that they can bring parties of clients along. Two big firms have booked the whole theatre for two evenings next year.' (*Reynolds News*, November 13, 1960.)

river and the best shooting for their companies. ... As a consequence, these amenities are now being priced far beyond the means
of the ordinary individual.'[1]

In evidence to the Royal Commission on Taxation (1952–5) the
Inland Revenue Staff Federation had this to say about expense
account developments:

> There is more commercial entertaining done today than ever
> before, much of it an extravagance and dissipation for which there
> is no real business necessity. ... The principal directions in which,
> apart from a general lack of economy in production and admini-
> stration, luxury or extravagant spending is in evidence are:
> (a) Advertising for purely goodwill purposes.
> (b) Motor cars bought and run ostensibly for business use, but
> very largely devoted to private use.
> (c) Commercial entertainment of all kinds.
> (d) Lavish expenses allowances to Directors and Executives.[2]

These observations were not dissimilar from many made in the
United States in recent years. 'Probably no aspect of our tax *mores*
has received more attention recently,' wrote *The Reporter* in 1959,
'than the expense account.'[3] It quoted an estimate published in the
Yale Law Journal that expense-account spending might be con-
servatively put at $5 billion a year resulting in a tax loss to the
Treasury of from $1 billion to $2 billion annually.[4]

These short notes on certain categories of benefits in kind do little
more than amplify the conclusion reached by the Royal Commission
that the provision of such benefits is capable of becoming an abuse
of the tax system. An exhaustive survey would have to take account
of many other categories.[5]

[1] *The Times*, January 19, 1962.

[2] *Volumes of Evidence, Royal Commission on Taxation 1952–5*, Document 243.

[3] *The Reporter*, April 16, 1959. See also Blake M. F., 'Current Trends in Fringe
Benefits', *The Journal of Accountancy*, September 1958.

[4] Rothschild V. H. and Sobernheim R., *Yale Law Journal*, July 1958. Compare
also an article by Mr G. J. Kienzle 'The Business of Christmas Giving' in *The
Reporter* (December 25, 1958) with an article in *The Times*, 'Christmas Gifts for
the Man who has Everything when Expense is no Object' (December 17, 1959)
and the 1958 guide to The Directors' and Executives' Exhibition in London.

[5] Such as: gifts of assets or their sale at less than market price; holiday expenses paid on business account; travel and other expenses for 'man-wife'
companies; private medical care for directors and executives; private sickness
benefits and premiums for accident insurance; legal aid for employees; club
subscriptions; telephone rentals; gardening and other domestic services; the
supply of newspapers and journals; subscriptions to professional, technical and
other societies; interest-free loans; 'living on the till'; suits and other articles of
clothing provided by employers; the computation of professional profits under
Schedule D; the provision of shooting, fishing, sports clubs and similar facilities;
free coal; consumer goods at free or reduced rates; professional school fees
schemes, and other items.

The nature and extent of such provision may be perfectly legal and yet, for reasons already mentioned, the benefits may not be fully reflected or reflected at all in the official statistics of pre-tax income. They may take a form analogous to splitting income in the family and similarly change over time the pattern of distribution. They may be associated with many other manipulative methods discussed in earlier chapters and consequently involve very difficult problems of valuation. That they now represent an important element in the standard of living for some groups in the population seems indisputable. As such, some forms of benefits may merge, almost imperceptibly, into the general problem of illegal avoidance—or evasion.

Avoidance and Evasion

Both Royal Commissions of 1920 and 1952–5 discussed the problem of evasion at length. Both were concerned as to its extent, and to the long lapse of time between the adoption and spread of a particular device and attempts by the legislature to nullify it. The material in Appendices D and E of this study indicates that in some instances the interval of time may be considerable. Other examples were cited by the Inland Revenue Staff Federation in its evidence to the 1952–5 Royal Commission, one involving a lapse of over twenty years.[1] This aspect of the matter is pertinent to our study, for although the extent of evasion may not be substantial today, nevertheless it may have been more so at certain periods since 1938. The statistics of income for such periods could thus have been affected in unknown ways.

Towards the end of the 1940's, for instance, there appears to have been considerable disquiet about the extent of evasion. In 1948 *The Economist* reported that 'tax evasion in the eyes of some Inspectors is no longer a risky proceeding so much as a standard trade practice. . . . Half a year's Revenue is awaiting collection if evasion could be detected and arrears vigorously pursued. . . .'[2] In the Appropriation Accounts for 1949–50 the Comptroller and Auditor-General referred to the report of a departmental committee which had 'confirmed the view that evasion is serious and widespread'. This committee had shown particular concern about the large number of self-employed people and small companies reporting little or no 'net true income'. The same questions were being voiced ten years later.[3]

---

[1] 'The well-known partnership device established by the case of *Osler v. Hall*, decided in 1933, still flourishes' (*Vols. of Evidence, Royal Commission on Taxation, 1952–5*, Document 243).

[2] *The Economist*, August 7, 1948.

[3] See various issues of *The Financial Times* for March and April 1960.

As part of the campaign at the end of the 1940's to tighten up on evasion 'the Board informed Inspectors of Taxes in June 1949 of the serious view they took of the extent of evasion . . . and urged them to take full advantage of various relaxations in procedure and other reliefs which had been authorized in recent years, so as to find as much further time for back duty work as possible. An intensified search for new liabilities has begun. . . .'[1]

The Inspectors themselves, through their Association, summed up the position, as they saw it early in the 1950's, in a Memorandum to the Royal Commission:

Legal avoidance has become a science with its own inventors and practitioners. The history of the Income Tax in the last thirty years is the story of a war between the experts who have devised schemes to enable their wealthy clients to reduce the burden of their tax and the legislature seeking constantly to frustrate them—a war fought out in endless and costly skirmishes in tax offices, and in battles and even campaigns in the Courts—a war in which again and again the taxpayer has won a temporary advantage only to be countered by fresh powers invoked by the legislature—until the web of the law is so tangled that only the experts on either side can unravel it. In almost every case the machinery of avoidance depends on a legal fiction—the separate personality of the Private Limited Company. At its simplest the taxpayer, whether trader or investor, steals an advantage over his less clever or more conscientious fellow by escaping surtax on undistributed profit; but he pays himself expenses and other allowances in money or in kind; he rewards himself for giving up the Direction of his own business or for undertaking not to compete with himself; he turns invested income into earned income by paying himself for managing his own investments; the property dealer converts his trade into a series of casual gains by forming a separate company to handle each transaction; the builder turns each speculative scheme into a casual gain by entering into a separate contract (at an unprofitable price) with a separate company which has only one transaction of buying and selling and then expires; the manufacturer buys a moribund company with large accumulated tax losses which he sets against his subsequent profits; and so on.[2]

The Memorandum ended by saying that 'avoidance of tax, whether legal or illegal, is reaching the proportions of a social evil'. Only two years before this was written, the Board had published its Report for 1949 which concluded that there had been a 'very considerable redistribution in incomes' since 1938 both before and after tax.[3]

[1] Memorandum of Evidence from the Inland Revenue Staff Federation, *Vols. of Evidence, Royal Commission on Taxation*, Document 243.
[2] *Vols. of Evidence, Royal Commission on Taxation*, Document 32.
[3] *BIR 92*, p. 86.

The Royal Commission, in examining witnesses, took up these allegations of avoidance. Giving evidence on behalf of the Association of Certified and Corporate Accountants in 1952, Mr J. E. Harris said that 'doubtful practices are not frowned upon today to the same extent as they would have been twenty-five years ago.'[1] Sir Alfred Road and Mr E. R. Brookes on behalf of the Board agreed, in answer to questions, that by 1948 'it was clear that there was a good deal of abuse of the expense allowance system, and the legislation was introduced in order to remedy that abuse'.[2] This legislation was the 1948 Finance Act to which reference has already been made.

A number of those who gave evidence on the problem of evasion drew attention to staff shortages in the Board, particularly in the Inspectorate. The prevention and punishment of evasion depended greatly on the efficiency of this branch of the Board's organization. The Inland Revenue Staff Federation, for example, after quoting a report from the Public Accounts Committee that the Board was understaffed,[3] pointed out that among 7,937 cases of evasion investigated between 1948 and 1951 (from which £4m. in penalties was imposed) there was not a single prosecution.[4] The Federation contrasted this treatment with what they described as 'the squalid cases of false claims for National Insurance and National Assistance benefits dragged into Court'.

Ten years later the Board was still faced with shortages of staff. 'The Inland Revenue,' reported *The Economist*, 'has fallen upon hard times: since the war it has steadily failed, and is still failing, to get anything like the number of tax inspectors it needs.'[5] That the public image of the Board's responsibilities had been seriously damaged by continuous attacks on taxation in general thus hampering recruitment was, according to some commentators, one factor in the situation. Another was said to be the loss of some senior men to

[1] Minutes of Evidence, July 4, 1952, *Vols. of Evidence, Royal Commission on Taxation*, 3479.

[2] *Ibid.*, July 8, 1952, 5655.

[3] *Third Report of the Public Accounts Committee, Session 1950–1*.

[4] *Vols. of Evidence, op. cit.*, Document 243, and *BIR 94*. See also the article by 'A Chartered Accountant' on 'Revenue Penalties and a Revolution in Back Duty', *British Tax Review*, July-August 1960, p. 285.

[5] *The Economist*, March 4, 1961, p. 842. *The Seventh Report of the Estimates Committee* (Session 1960–1), published some months later, wholly devoted to the work of the Inland Revenue, confirmed this statement about staff shortages. 'It is, therefore,' said the Committee, 'all the more serious to learn from evidence given by the Chairman of the Board that the Inspectorate is about 300 below its authorized strength. Despite the increases both in yield and complexity of Income Tax and in numbers of taxpayers there are at the moment only about 1,900 fully-trained Inspectors compared with about 1,750 before the war' (July 1961, p. x).

industry whose expert knowledge of the tax system was clearly of value to their new employers.[1]

The attempts in the Finance Acts of 1960 and 1961 to counter various types of avoidance and evasion brought to the fore again these questions of staffing. It is not necessary to pursue them here for the whole subject of staff recruitment and training, penalties, prosecutions and so forth lies outside the main interests of this study. Nevertheless, it is impossible not to express sympathy for an over-worked staff, struggling to maintain good relations with the taxpayer at a time when many of them must have been conscious of a public opinion which regarded their responsibilities as an anti-social job. The criticisms that we have made about the Board's statistics must not, therefore, be taken as an attack on those who man the Department. They are directed at that body of opinion which has simultaneously moulded the public image of the Board's work and denied it the resources to discharge its responsibilities efficiently.

The only purpose of this brief excursion into a variety of topics associated with the development of fiscal law and its administration since 1938 has been to underline their relevance to the study of changes in the distribution of personal incomes. The general neglect of these topics by other students of income distribution gives added point to this approach. But while evidence may be brought together suggestive of trends and shifts in opinion and practices, it is far more difficult to be in any remote sense precise about the influence of these many factors on the general erosion of the tax base since 1938. Legal avoidance through and as a result of a whole complex of cause and effect has undoubtedly grown substantially since 1938 as earlier chapters have indicated. According to the material examined, evasion would seem, on the other hand, to have been more prevalent at some periods than at others. What effect both factors in association with other forces at work have had on the distribution statistics it is impossible to estimate with any assurance.

Mr Prest, in his book on Public Finance (1960), came to the conclusion that 'it would certainly seem to be the case that disguised income in the form of expense allowances is more common in the United Kingdom than it was in pre-war days, and probably (though not certainly) the under-reporting of higher incomes is relatively greater in this respect than lower incomes'.[2] In examining the Board's statistics, he compared the number of pre-tax incomes in 1938 and 1955 at £2,000 and over and allowed for a two-and-a-half fold rise in personal money income per head in the United Kingdom between these years. The results of this exercise, he said, 'do suggest that the upper income groups in the United Kingdom may have

---

[1] *The Observer*, June 18, 1961.
[2] Prest A. R., *op. cit.*, p. 261.

learned something about expense allowances, which perhaps Americans knew at an earlier date'.

In Appendix F, which illustrates Mr Prest's approach, we have compared the number of pre-tax statutory incomes by ranges in 1938 and 1958. Allowing for a threefold rise in personal money income per head during these twenty years, and putting on one side momentarily all our criticisms of these data, the exercise shows a remarkable fall in the number of pre-tax incomes in 1958 corresponding in value to the range £2,000 and over in 1938. At the level of £6,000 and over before the war, the number of incomes fell from 19,024 to 4,125 in 1958.

This extraordinary behaviour in the statutory world of income of a relatively small group of top pre-tax incomes contributed substantially to what has been called the income-wealth revolution in the United Kingdom since 1938. Such behaviour is, however, better expressed in the notation of statistical illusion than in the language of equality.

CHAPTER 9

The Need for a New Approach

I

TO study the rich and the sources of power in society is not the kind of activity which comes easily to social workers attempting to understand the human condition. Traditionally, they have been concerned with the poor and the consequences of poverty and physical handicap. They have thus tended to take—perhaps were compelled to take—a limited view of what constituted poverty. It was a view circumscribed by the immediate, the obvious and the material; a conception of need shaped by the urgencies of life daily confronting those they were seeking to help. Insofar as they looked at relativities and inequalities in society—which they seldom did—they restricted their studies to the day-to-day differences in levels of expenditure on the more obvious or more blatant necessities of life. Daily subsistence was both the yardstick and the objective.

Far-reaching changes affecting the structure and functions of social institutions; general improvements in material standards of living; and the growth of knowledge about the causes and consequences of social ills in the modern community, are now forcing on us the task of re-defining poverty. Subsistence is no longer thought to be a scientifically meaningful or politically constructive notion.[1] We are thus having to place the concept of poverty in the context of social change and interpret it in relation to the growth of more complex and specialized institutions of power, authority and privilege. We cannot, in other words, delineate the new frontiers of poverty unless we take account of the changing agents and characteristics of inequality. How then is poverty to be measured today and on what criteria, secular, social and psychological?

Each generation has to undertake anew this task of re-interpretation if it wishes to uphold its claim to share in the constant

[1] For an analysis of recent changes in the interpretation of 'subsistence' see Lynes Tony, *National Assistance and National Prosperity*, Occasional Papers on Social Administration, No. 5, 1962.

renewal of civilized values. Yet the present generation, it must be conceded, has been somewhat tardy in accepting this obligation. It has been too content to use the tools which were forged in the past for measuring poverty and inequality.

These tools are now too blunt, insensitive and inadequate. They do not go deep enough. These also are the lessons thrown up by this particular study of one of the primary sources of knowledge about the distribution of incomes. They yield a surface view of society which is increasingly at variance with other facts and with the evidence of one's eyes.

What, it may be asked, has prevented us from sharpening these tools of inquiry and applying them with more precision to contemporary Britain? Three by no means inclusive reasons may be provocatively advanced.

First it may be said that modern societies with a strongly rooted and relatively rigid class structure do not take kindly to self-examination. The major stimulus to social inquest in Britain during the present century has come from the experience of war. On each occasion this experience was sufficiently mortifying to weaken temporarily the forces of inertia and resistance to change. In the absence of such stimuli in the future, we may have quite consciously to invent and nourish new ways and means of national self-examination. This task may be harder to discharge in the face of rising standards of living and the growing influence of the mass media of complacency. Perhaps the most powerful challenge of all will come from the notion that economic growth will solve all our social problems involving choice, distribution and priorities.

Secondly, it becomes clearer as we learn to distinguish between the promise of social legislation and its performance that the present generation has been mesmerised by the language of 'The Welfare State'. It was assumed too readily after 1948 that all the answers had been found to the problems of health, education, social welfare and housing, and that what was little more than an administrative tidying-up of social security provisions represented a social revolution. The origins and strength of this climate of opinion—some illustrations of which are given later—will no doubt continue to puzzle historians for a long time to come.

Thirdly, and concomitantly, the 1950's saw the spread of the idea that some natural built-in 'law' was steadily leading to a greater equality of incomes and standards of living. It followed implicitly from this theory that further economic growth would hasten the operation of the natural law of equalization. This was not a new thesis; Marshall had stated it hopefully over seventy years ago: 'the social and economic forces already at work are changing the distribution of (income) for the better; ... they are persistent and increasing in strength; and . . . their influence is for the greater part cumulative'.[1]

This thesis received powerful support from Government and independent statisticians in the 1950's, notably in the studies made by the Board of Inland Revenue and by Mr Lydall and Professor Paish which have been extensively quoted here. These essays in income distribution were the active agents in encouraging this particular examination of the basic statistics. They were the point of departure for the present author. But what seemed three years ago a fairly simple matter of re-assessing the evidence became, as the work went on, a complicated problem of arrangement and communication. This had better be explained.

II

We began with a set of specific questions about the validity of the official statistics on the distribution of incomes since 1938. As we got more deeply involved with the fundamentals of definition, measurement and analysis we still tried to keep these questions as the central focus of the study. Hence we made no attempt to quantify, in national aggregates, changes in the distribution of incomes over the twenty years. Indeed, it soon became apparent that as we moved outside the confines of statutory income this was an impossible task without far more research and, above all, more basic data. Nevertheless, in order to dissect and demonstrate the limitations of the existing data, it was necessary to accumulate a mass of supporting material which, narrowly considered, may seem to the observer to be only of marginal interest.

This material, as it was assembled and analysed, began to show the dimensions of more fundamental problems. We were led away, so to speak, from the severely statistical short-run approach to a consideration of long-run forces affecting the economic and social structure of British society. We began to see that many of the processes of rearranging, transforming and spreading income-wealth were no simple manifestations of high taxation since the 1940's. They were not ephemeral mutations, to be eradicated by such acts as raising the surtax level, but evidence of deeply-rooted trends in the economic structure of a changing society. There emerged, as we looked at some of the causal factors, what seemed to us to be highly significant connections between apparently unrelated social and economic institutions. We saw, for instance, the law on the alienation of income reflecting long-run changes in family

[1] Marshall A., *Principles of Economics*, 8th ed., 1948, p. 712. The word 'income' is substituted for 'wealth' in the original quotation, for Marshall uses them almost interchangeably in this part of his book.

relationships, the points at which the private education sector connected with the shifting concept of 'charity', and the rise of new institutions for storing and redistributing income and property rights. The growth of new social mechanisms for exercising and extending command over resources-in-time was seen as one of the links between these different institutions.

These observations, which took us a long way from the comparatively simple task of identifying sources of statistical bias in the official data, are not restricted in time. They are as relevant to the real world of income-wealth in the 1960's as they are to the 1950's. In this sense, the perspective is very different from that which enclosed the origins of the study—a cross-examination of the statistics for the period 1938–58.

It may be said, therefore, that much of the material presented here, though necessary as supporting evidence for the central argument, has a bearing on the more fundamental problem of what constitutes inequality and what poverty means in the second half of the twentieth century. To have pursued these themes would have taken us far beyond the self-imposed limits of this study. What we have shown, so far as the original aims of the study are concerned, is how little we know about some of the measurable characteristics in the changing equation of inequality. In this respect we have thus been singularly unconstructive. But perhaps, unlike the young, we cannot begin to learn until we consciously admit to our own ignorance.

III

From a practical point of view, therefore, this study is unlikely to be immediately helpful to those who attempt the task of estimating from the official statistics changes in the distribution of personal incomes. Though we have criticized them at length we must, however, recognize that they are sometimes placed in positions in the real world of action of making the best of imperfect statistical materials. They are confronted with the data; they know that the Board of Inland Revenue has sought to make improvements by undertaking sample surveys, and they may note the absence for many years of any serious criticism of this body of knowledge. Pressed for answers and often pressed for time they develop such techniques as are available to guard against erroneous conclusions.

But the imperfections accumulate. They stem from changes taking place in society at large; from more different and complex ways of getting, spending, spreading and storing income; from social and demographic forces which change the characteristics of income units, families and economic classes; from a process of fission affecting the nucleus of rights which we call property resulting in larger concentrations of corporate wealth and power without ownership; from new forms of social benefit and privileged consumption for the family as well as the individual that flow from achieved and inherited status rather than contract; and from the growth of fiscal legislation providing special concessions and distinctive privileges in favour of certain groups or classes in the interests of economic efficiency, saving and investment, professional self-advancement and other socially approved objectives.

Previous chapters have traced some of these complex factors at work in society and have indicated the need for a new approach to the problem of measuring inequalities in the distribution of income and wealth. They have shown that the present mould into which the statistics are cast is increasingly delusive as a frame for knowledge. We are in fact expecting too much from the crumbs that fall from the conventional tables. We cannot so easily reduce the complex to the simple. By attempting to do so we ignore age, sex, kinship, occupation, civil status and other critical variables, and we lose sight of the importance of defining clearly the unit of income, the unit of time and the concept of income.

Some of the criticisms we have made about the statistics of income are also applicable to the statistics of wealth. For example, as long ago as 1939 Mr (now Sir) H. Campion suggested that the distribution of property ought to be measured by families.[1] He drew attention to the tendency for property and the benefits arising therefrom to be shared with wives, children and other relations. This trend, he considered, could account in part for the reduced inequality of the distribution of property since before 1914. More recent studies by Mr Revell and Messrs Lydall and Tipping emphasize the importance of this factor. The latter commented on the 'growing tendency for owners of large properties to distribute their assets amongst the members of their families well in advance of death'.[2]

The evidence assembled in this study confirms this change in the pattern of economic relationships within the family. The causes of this change are probably associated in some measure with changes in family life and in the status accorded to wives, children and other kin relations amongst the wealthier classes—a subject little examined by sociologists because of the difficulties of access to the primary data. Studies of the distribution of income and wealth which show a gentle decline in inequality in Britain in terms of persons or income units may, therefore, be concealing a significant growth in inequality between families. Professor Lampman's investigation into the changes in the share of wealth held by the top wealth-holders in the USA concluded that the decline in inequality between 1922–56 had

[1] Campion H., *Public and Private Property in Great Britain*, 1939, p. 110.
[2] Lydall H. F. and Tipping D. G., *op. cit.*, p. 85.

been overstated on an individual compared with a family basis.[1]

To what extent then, it may be asked, are we becoming prisoners of the statistical houses we built in the past to accommodate the social data of that age? The appeal of continuity in the analysis and presentation of large masses of data must always be strong in a historically conscious society, and statisticians (like other people) have their own particular reasons for not wishing to change. Their preferences for stability and order are reinforced when those who use the results of their work fail to appraise its relevance to a different social structure and to changes in economic and institutional relationships.

Though we may now be collecting more information in the field of income distribution than we were twenty or thirty years ago it does not follow that we are better informed about the essential constituents; when or how they may be added together, arranged and presented, and what conclusions are valid. In many respects, the standard of critical thought brought to bear by the users of these statistics has not kept pace with either the volume and complexity of the facts themselves or with the development of methods of statistical analysis. Supported intellectually with more sophisticated aids, we seem to have been entering for races before we are able to walk.

Much of this running in the field of income distribution has been over ground heavily sown with 'routine administrative statistics'. It is not necessarily a criticism of the Board of Inland Revenue to say that its annual reports are largely a record of 'work done'. Consequently, its definitions, insofar as they are needed for reporting on the discharge of responsibilities, tend to be administrative definitions. The Board may justly reply in such terms to the comments we have made. It is no fault of the Board, it may be said, if its statistics of 'work done' are misused or misinterpreted by others. All this is undeniable. On the other hand, it should be pointed out that the Board does not reprove those who mishandle its statistics. Every year, on the occasion of the publication of the Board's report, the national press is filled with accounts of how inequality is diminishing and how the rich are disappearing as a distinctive class. It is seldom if ever announced that in most years these accounts are based on the collection and tabulation of administrative statistics. In time we come to believe them as these results are purveyed without qualification, and more and more textbooks on public finance and allied subjects report them as established, indisputable facts.

Nor is the Board entirely free from criticism in other respects. In

[1] Lampman R. J., *Review of Economic Statistics*, 41, November 1959, pp. 379–92. See also his Study Paper No. 12, *The Low Income Population and Economic Growth*, prepared for the Joint Economic Committee, United States Congress, December 1959.

response, no doubt, to demands over the last twenty years from income statisticians, the Central Statistical Office and other quarters, the Board has attempted to convert or adapt its administrative statistics for use as indicators of change in the distribution of incomes. These attempts have been pursued since 1938 without any adequate discussion of the fundamental problems of defining the units that are tabulated. In its 1949 Report, as we have already pointed out, the Board presented an analysis of its administrative statistics, accompanied by two charts, which purported to show, in the Board's words, that 'there has been a very considerable redistribution in incomes since pre-war'.[1] There was no suggestion here that it was simply reporting on 'work done'; on incomes assessed and taxes collected under various schedules. The data were presented in 1950 as a contribution to the study of income distribution. As the Board must have expected, they were widely and uncritically consumed in this sense.

Their general acceptance is perhaps but one of many symptoms of the intellectual narrowness of writers on public finance and related subjects. They seem to be oblivious of the work being done in neighbouring fields by demographers, sociologists, lawyers, accountants and others. Since 1945, for instance, there has been a veritable spate of literature on what is described as 'tax planning', to say nothing of the rise of new 'professional' groups of tax consultants, pension consultants, death duty consultants, and 'one-man company' experts. These phenomena suggest that the conventional models of economic man and income units may be wearing a little thin in terms of fiscal statistics. It is a defensible thesis to advance, against the background of the social changes we have sketched in, that 'yearly income' (in the traditional sense of disposable cash) is now less important among the top income-wealth groups in society. Long-run developments in family relationships and fiscal law; the shifting alignment of forces controlling access to a power system as distinct from individual property rights; the transformation of individual benefits in cash into family benefits in kind; the changing individual age cycle of earning and non-earning; the spreading and splitting of 'income' over life and over the lives of several generations; the metamorphosis of income into capital; the growth of rewards in the form of tax-free payments and the use of credit facilities; all these make 'cash in hand' less necessary for the business of daily life for certain classes and living on overdrafts, trusts and other forms of command-over-resources more fiscally rewarding if, at times, perhaps a little irksome.

While it is easy, of course, to make fun of any specialist group for

---

[1] *BIR 92*, p. 86.

N

their unalloyed absorption in a special branch of knowledge, it must
be admitted that they are not suffering from a particular occupational
disease but from a general epidemic of intellectual short-sightedness.
We are all, in some senses, victims and offenders, confronted as we
are with the problem of the growth of knowledge in complex
societies and the division of intellectual labour into smaller and
smaller segments.

Some of these handicaps might be reduced if the statistics of
income and wealth were presented in a broader and more analytical
framework. Not only do we need to know much more about what is
being measured, but alternative ways of presenting the results are
required as a corrective to the conventional 'snapshot' view of
statutory income. Much too could be learnt if these different analyses
were tabulated in terms of age, sex, civil status, family relationships,
occupation, types of income, categories of expenditure on certain
forms of goods and services, and allocations to savings of various
kinds. Some of this information could be cross-checked and amplified
by the use of national sample budget surveys of expenditure, saving
and the ownership of property. A national census of all discretionary
trusts, family settlements and covenants would be essential for the
purposes of both income and wealth distributions, and the Government would have to take a number of steps to compel insurance
companies, charities and corporate bodies in general to furnish
much more information than they do at present.

Given the will to be a less secretive society, there is nothing im-
practicable about these suggestions. They would, of course, take
time and resources to implement but the gain in terms of public
enlightenment about the constituents of equity could be impressive.
It would no longer be possible for pronouncements to be made of
such a general character as the 1950's endured. We may consider, as
a conclusion to and a commentary on the material assembled in
earlier chapters, a few of the statements made in recent years about
changes in the distribution of income and wealth.

IV

The Board's report on its 1949 statistics was the text for many
writers in the early 1950's. It provided, for example, part of the
factual basis for Mr de Jouvenel's influential study *The Ethics of
Redistribution* published in 1951. *The Economist*, in a series of
articles on 'Personal Incomes', went further than the Board in
claiming that its figures indicated 'a vast redistribution of incomes'
since 1938.[1] Two books by Messrs. Lewis and Maude, which came
in for a great deal of attention at this time, were clearly stimulated
by official statistics depicting a considerable decline in inequality.

[1] *The Economist*, February 4, 1950, p. 248.

The first, which the authors thought of calling 'The Decline and Fall of the Middle Classes', appeared in 1950. The second, *Professional People*, in which the conclusion was reached that 'the professions have tended to come off worse than any other section of the community except retired pensioners', was published two years later. Mr Colin Clark's *Welfare and Taxation*, which followed in 1954, accepted the Board's statistics of incomes without qualification.

These and similar studies contributed to the shaping of public opinion and furthered the acceptance of the Board's statistics. In more specific ways, they were significant in the formation of policies. Thus we find, in 1952 and again in 1954, the Board's comparison of changes in the distribution of incomes between 1937–8 and 1949–50 being submitted, 'as adequate data', to Mr Justice Danckwerts as adjudicator on the remuneration of general practitioners and to the Royal Commission on the Civil Service.[1] Those responsible for making recommendations to the Government must have found it difficult, guided as they were by little expertise on these fiscal matters, to refute or qualify official statistics of this order of complexity.

By the middle 1950's these conclusions had become part of the history of our times. Professor Robbins, reflecting on trends in the distribution of incomes and wealth, wrote of a tax structure which 'relentlessly, year by year, is pushing us towards collectivism and propertyless uniformity'.[2] In much the same strain, Mr Brendon Sewill and the Bow Group interpreted the Board's statistics as depicting 'the virtual extinction of the pre-war "surtax class"'.[3] They were also employed by Mr Enoch Powell in his book on saving to show that the trend towards equality, apparent from the official data for 1938 to 1949, continued to operate for the later years to 1958.[4]

All these writers who attempted comparisons over time in the distribution of income and wealth took as their baseline the official statistics for 1936–8. We need not recapitulate here the criticisms we have already made about the validity of many of these data. Whatever their defects, however, the pattern of inequalities which they disclosed was the one which Keynes, despite his belief in the value of a significant range of inequality, attacked as socially unjust for its magnitude.[5] Professor Jewkes, also voicing his disquiet at this

[1] See 'Memorandum on Changes in the Value of Money and Standards of Living 1938–51' by Professor R. G. D. Allen submitted by the British Medical Association and Appendix II and III, *Minutes of Evidence, Royal Commission on the Civil Service*, Day 7, 1954.

[2] Robbins L., *Lloyds Bank Review*, October 1955, p. 18.

[3] Sewill B., *op. cit.*, p. 10 and *Taxes for Today*, Bow Group, 1958, p. 27.

[4] Powell E., *Saving in a Free Society*, 1960, p. 127.

[5] Keynes J. Maynard, *General Theory of Employment, Interest and Money*, 1936, p. 374.

time, asked the question: 'must we be prepared to tolerate, so long as we had a system of private enterprise, the apparent paradox that although men were equal politically, they must inevitably remain as unequal economically as they now are?'[1] Twenty years later, and after one of the greatest wars in British history which might have been expected to exert a profound influence on the distribution of wealth, these measures of inequality in the 1930's were being regarded as acceptable criteria for determining the degree of inequality that might be socially justified.

These views were strengthened by the effects on the tax structure of all the complex forces we have analysed which cumulatively were enlarging the dustbowl of taxable capacity. 'We have travelled far down the vicious path of decadent tax systems,' said Mr Kaldor at the end of his book *An Expenditure Tax*, 'the path of charging more and more on less and less.'[2] This journey has, moreover, been hastened by the growth of social welfare provisions in fiscal law; dependency benefits for children and other family members, allowances for those undertaking higher education and 'self-improvement', relief and exemption in old age, and deductions and discriminatory reliefs for particular categories of saving for retirement and other purposes by certain sections of the population. While it is outside the scope of this study to examine the problems of equity raised by the growing use of the fiscal system as a vehicle of government action in the social policy field, it should be pointed out that this development contributes to the further erosion of the tax base. 'A programme of tax allowances which costs the State in the neighbourhood of £450m. in lost revenues deserves to be recognized,' said Professor Cartter in referring to dependants' allowances in 1952, 'as a major element in the programme of social welfare.'[3]

Taken together, all these factors tend to give the standard rates of taxation a spurious role in public life. They involve, as these rates become increasingly ineffective for some groups but not for others, a subtle kind of moral dishonesty. We are thus led to claim, with varying degrees of pride or shame, that Britain is the most highly taxed nation in the world.[4] Or that progression in the tax structure is steeper than it has ever been.[5] Or that this nominal structure

---

[1] Jewkes J., in Introduction to Campion H., *Public and Private Property in Great Britain*, 1939, p. xi.

[2] Kaldor N., *op. cit.*, p. 242.

[3] Cartter A. M., *Population Studies*, March 1953, Vol. VI, No. 3, p. 227.

[4] See, for example, Prof. Alan Peacock in *The Welfare Society*, Unservile State Papers, No. 2, 1960, p. 6. It is doubtful whether international comparisons have much validity without far more research than has hitherto been attempted into the problems of definition and the actual working of fiscal systems.

[5] When income tax, national insurance and health contributions, and local rates are combined, there appears to be little progression for most employed

justifies great increases in salaries and tax-free rewards for certain sections of the population—an argument put into effect increasingly in the 1950's.[1] The best of both worlds is thus achieved by some at the expense of others as our tax systems move further away from the notion of equitable neutrality in relation to the dependency burdens of kinship, the costs of higher education and self-advancement, the saving and spreading of income and its transformation into capital, the consumption of particular categories of goods and services and so forth.

Future historians with access to the public archives may find a fascinating theme to unravel in studying the connections over the past twenty years between the erosion of the tax base brought about by the many forces discussed here, changes in redistributive social policies, reductions in the 'burden' of income tax and surtax that took place after 1951, and the search by governments for new and more effective fiscal systems. They will note the decline in favour of the progressive tax system along with the principle that ability to pay is the right democratic criterion. They will no doubt also analyse the attempts to give greater weight to regressive systems and taxes on consumption and, in particular, the shifting of Exchequer burdens to local rates. Above all, they will be led to study the fashioning of a new and massive fiscal system in the shape of national insurance and health service contributions. By 1961–2 these two regressive poll-taxes were yielding close on £1,200m. a year—or six times as much as the yield from surtax before the surtax level was raised to £5,000. 'Contributions' levied without regard to personal circumstances, were by then the most important form of direct taxation experienced by something like half the working population. For the Treasury in particular, they had the advantages of simplicity in administration, the prospect of a steadily growing yield, and a way of redistributing part of the costs and consequences of tax avoidance and erosion.

The implications of this policy could, however, be quite profound. The ease with which this new system can be deployed, and its comparative freedom from avoidance devices, may mean in the long run a gradual abandonment of the search for equity. A situation is created in which it becomes psychologically harder to press for reforms in the older form of direct taxation. Action appears to be less urgent if revenue can be effectively raised in other ways. For many years now, *The Economist* and other commentators have

persons with earned incomes of under £2,000 a year. This point was made by Mr G. Fox in 'New Taxes for Old', *Crossbow*, Spring 1961, p. 46.

[1] See, for example, Copeman G., *Promotion and Pay for Executives*, 1957, and 'The Middle Classes 1951–8' by the Financial Editor of *The Financial Times*, November 14, 1958.

argued that the only solution to the problem of avoidance and erosion—perhaps democracy's hardest domestic problem—is greatly to reduce 'the burden' of the progressive tax.[1] In other words, the only answer to the challenge of moral behaviour is—in the ultimate analysis—to abolish the need to be moral.

V

These general reflections on the lessons of our study have led us far from its point of departure. That was to examine the statistics on the distribution of incomes before tax and to inquire into the validity of the conclusions reached by others who had studied these data. Whatever else may be said about the criticisms we have made about the sources of information and about those who have interpreted the material, there can be little dispute with the conclusion that we know less about the economic and social structure of our society than we thought we did.

It follows from this that we should be much more hesitant in suggesting that any equalizing forces at work in Britain since 1938 can be promoted to the status of a 'natural law' and projected into the future. As we have shown, there are other forces, deeply rooted in the social structure and fed by many complex institutional factors inherent in large-scale economies, operating in reverse directions. Some of the more critical of these factors, closely linked with the distribution of power, and containing within themselves the seeds of long-lasting effects—as, for instance, in the case of settlements and trusts—function as concealed multipliers of inequality. They are not measured at present by the statistics of income and only marginally by the statistics of wealth. Even so, there is more than a hint from a number of studies that income inequality has been increasing since 1949 whilst the ownership of wealth, which is far more highly concentrated in the United Kingdom than in the United States, has probably become still more unequal and, in terms of family ownership, possibly strikingly more unequal, in recent years.[2]

It has not, however, been the purpose of this study to estimate the total effect of all the forces which determine the class distribution of incomes and wealth. We have simply attempted to show that fact and economic theory are at variance, and that no conclusion which takes account of an ageless individual and forgets the family, which measures 'income' and omits 'wealth', which disregards the unit of time in command-over-resources, which fails to inquire into the

[1] See, for example, *The Economist*, April 30, 1960.
[2] See earlier references to the studies by Mr Brittain, Messrs Lydall and Tipping, Mr Revell and Professor Lampman, and Mr Douglas Jay's *Socialism and the New Society*, 1962, especially pp. 191–215.

meaning of power, which avoids investigating the interlocking connections between social and economic institutions, and which is oblivious of the key role now played by the educational system in the social distribution of 'life chances', can be relied upon in the context of the social changes we have depicted. Ancient inequalities have assumed new and more subtle forms; conventional categories are no longer adequate for the task of measuring them.

APPENDIX A

Questionnaire Submitted to the Board of Inland Revenue in March 1961

1. Who were required to make a return of income for tax purposes in the following years? Please also state:
   (*a*) at what ages individuals had to start making returns
   (*b*) how returns for married women and the retired were handled
   (*c*) in what circumstances were returns required at intervals of more than a year.

                              1936–7
                              1937–8
                              1949–50
                              1954–5
                              1955–6
                              1956–7
                              1957–8
                              1958–9
                              1959–60

2. Has there been since 1936–7 any legal obligation to include non-taxable income (including lump sum payments on retirement and in other circumstances) in a return?

3. Is it correct to say that, except in the case of the three special surveys, the classification of incomes tables (e.g. tables 57/8 in 101 Report):
   (*a*) are not based on any aggregation of income charged on one or other of the five schedules
   (*b*) exclude all earnings or income of wives not reported on husbands' returns
   (*c*) exclude unreported income from dividends and interest taxed at source.

4. How are the incomes of individuals entering and leaving employment and who earn for less than a year treated in the classification of incomes tables? If they appear as income units (providing their income is above the exemption limits) can any estimates be given of the numbers involved for 1937–8, 1949–50, 1954–5 and 1957–8 and by sex, age and marital status?

5. Can any information be given as to how the 1937–8, 1949–50 and 1954–5 samples were selected? Was account taken of age, sex, marital status and type of income?

6. Is Schedule C income excluded from all the classification of income by ranges tables? If so, can any estimates be given of its distribution by range of income, sex and age for 1937–8, 1949–50, 1954–5 and 1957–8?

7. How are back year assessments treated in the classification of income by ranges tables? Can any estimates be given of the amounts involved by ranges of income for 1937–8, 1949–50, 1954–5 and 1957–8?

8. How does the operation of Double Taxation Relief affect the statistics in the classification of income by ranges tables? Does income before tax include all types of overseas income before deduction of overseas tax?

9. How are trading losses dealt with in the classification of income by

ranges tables? Can any estimates be given of the adjustments made to the tables for 1937–8, 1949–50, 1954–5 and 1957–8 and by year of loss, age and sex?

10. The statistics of total income before tax are different in different tables of the Board's Reports, e.g. tables 21 and 56 of the 100th Report. Could a reconciliation be provided of these two sets of statistics for 1937–8, 1949–50, 1954–5, 1957–8 and 1959–60 with separate estimates for different assessment years?

11. As regards the tables relating to surtax since 1936–7 can any estimates be given for each year of the number of income units who fell into the following categories:

(a) children aged under 21
(b) widows aged over 60
(c) wives assessed under Schedule D
(d) men assessed under Schedule D

12. Where the income of a company is subject to a surtax direction, is it treated as income of the shareholders in the classification of income by ranges tables (a) in the year to which the surtax direction relates or (b) in the year in which it is distributed to the shareholders? Can any estimates be given of the number of such directions, classified by amounts of income involved and by income ranges of the shareholders concerned, and the total amount of income and tax involved, for 1937–8, 1949–50, 1954–5 and 1957–8?

13. What was the deficiency before tax in 1937–8, 1949–50, 1954–5 and 1957–8 attributable to unreported income from dividends and interest taxed at source by (a) men (b) women?

14. Is all interest paid (to banks, hire purchase and finance companies, persons and institutions but excluding mortgage interest) deducted before arriving at income before tax? Can any figures be given of the number of cases and the amounts involved for all types of interest deductions (excluding mortgage interest) in 1937–8, 1949–50, 1954–5, 1957–8 and 1959–60 and by income range, sex, age and marital status?

15. Can any figures be given of mortgage interest deductions (number of cases and amounts involved) in 1937–8, 1949–50, 1954–5, 1957–8 and 1959–60 and by income range, sex, age and marital status?

16. Can any figures be given of maintenance allowance deductions under Schedule A (number of cases and amounts involved) in 1937–8, 1949–50, 1954–5, 1957–8 and 1959–60 and by income range, sex, age and marital status?

17. Can any estimates be given (and on what basis) of unreported investment income attributable to accumulation of income under settlements and covenants during years of minority which counts as income neither of settlor nor life tenant nor beneficiaries for the years 1937–8, 1949–50, 1954–5, 1957–8 and 1959–60 and by income range of the settlor?

18. Do the tables classifying income since 1936–7 (e.g. tables 21, 57, 58, 59, 60, 61, 64, 65, 66, 67, 68 in 101 Report) relate to individuals or to individuals and married couples treated as one income?

19. How are separately assessed wives treated in these tables? Can any statistics be made available for the years specified in Question 1 showing

the number, income range, age and type of income of separately assessed wives?

20. Do these tables count the same individual in two income units in the case of (a) women in the year of marriage (b) widows and widowers in the year of the spouse's death? Can any estimates be given of the numbers involved in 1937–8, 1949–50, 1954–5, 1957–8 and 1959–60 and by income range, sex, age, and type of income?

21. Under what circumstances are married couples living apart treated in these tables as separate individuals? Are they automatically treated as separate if (a) a maintenance order or (b) deed of separation is in force? From what date are they treated as separated in the case of (a) divorce, (b) maintenance order, (c) deed of separation, (d) de facto separation? What evidence of de facto separation is required?

22. Can any estimates be given of the number of 'separated individuals' appearing in the tables for 1937–8, 1949–50, 1954–5, 1957–8 and 1959–60 by income range, sex, age and type of income?

23. How are the deceased treated in these tables in respect of that part of the year in which they were income receivers? If they are counted as income units for part of a year, can any estimates be given of the numbers involved in 1937–8, 1949–50, 1954–5, 1957–8 and 1959–60 and by income range, sex, age, marital status and type of income?

24. In the definition of 'income before tax' (101 Report, p. 69) reference is made to 'allowable expenses . . . and similar annual payments'. Are these defined by statute?

25. Does 'income before tax' exclude income in the form of sickness benefit paid by an employer, by an insurance company or similar scheme paying cash benefits or medical bills?

26. Have any statistics been published since the Finance Act 1948 (or can any be made available) showing the classification by ranges of taxable benefits in kind?

27. Can a list be given of the types of benefit in kind which were in 1937–8 and 1959–60 (a) taxable and (b) non-taxable?

28. In what circumstances has the income of children (whether above or below exemption limits) been aggregated with that of their parents in the classification of incomes by ranges tables since 1936–7? Can any estimates be given for 1937–8, 1949–50, 1954–5, 1957–8 and 1959–60 of the number of children whose total income was (a) aggregated (b) treated separately in these tables?

29. If child allowances are not claimed how does the Board ascertain the incomes of children?

30. How are the following sources of income for children aged under 21 treated in the classification of incomes by ranges tables each year since 1936–7?

    (a) local education grants

    (b) scholarships and bursaries from public and private bodies

    (c) the payment of school fees under covenants, settlements and by employers

    (d) school fees paid by means of endowment policies effected by employers (*e*) benefits in kind derived from the fathers' employment.

31. Can any estimates be given of the total amount of income before tax of such children in each category (*a*) to (*e*) for each year since 1936–7?

32. Were wives' earnings (with their different definition of net earned income) included as separate units in the main classification tables? For example: tables 63 and 60 in 101 Report.

33. Have any statistics been published since 1936–7 of 'missing earning wives' and all earning wives by ranges and type of total net income?

34. What was the deficiency in 1937–8, 1949–50 and 1954–5 before and after tax and by range of husbands' income in the reported earnings of wives?

35. How are the earnings of wives under Schedules A, B, C and D treated in the main classification of incomes by ranges tables? If any earnings or incomes are separately assessed are they then added for tabulation purposes to the incomes of husbands?

36. In the 101 Report the Board stated that 'no loss of tax was involved' as a result of husbands omitting or understating wives' earnings (p. 70). Presumably this means (*a*) that PAYE was deducted on the wives' earnings at a rate which took account of the husbands' income and (*b*) that the 1,500,000 wives involved assessed under Schedule E appear as single women in the statistics. If this is so, can any estimates be given for 1949–50, 1954–5, 1957–8 and 1959–60 of the distribution of wives' earnings by range of income for (*a*) these wives and (*b*) their husbands?

37. Can separate estimates be given of tax forgone on account of:
(*a*) child allowances
(*b*) dependent relatives
(*c*) housekeeper allowance
(*d*) age relief and allowances
(*e*) life assurance distinguishing between—
   (i) reliefs to individuals
   (ii) expenses allowances to employers in respect of premiums paid
(*f*) payments for retirement benefits distinguishing between—
   (i) allowances to employees and others
   (ii) expenses allowances to employers in respect of contributions and premiums
(*g*) payments for retirement annuities under S.22 of the Finance Act 1956
(*h*) National Insurance contributions distinguishing between payments by employers and insured persons for the years—

$$\left.\begin{array}{l} 1937\text{–}8 \\ 1949\text{–}50 \\ 1954\text{–}5 \\ 1957\text{–}8 \\ 1959\text{–}60 \end{array}\right\} \quad \text{excluding } (g)$$

38. How has the income of trustees been treated in the classification of incomes by ranges tables each year since 1936–7; particularly where the income is on trust for the benefit of an infant?

39. Are formal declarations now required from all makers of covenants in favour of family members to the effect that no agreement or under-

standing exists for the return, direct or indirect, of any part of the benefit? If so, please state the annual number of declarations deposited with the Board for 1957–8, 1958–9 and 1959–60.

40. Have any declarations during these years not been honoured? What steps are taken by the Board to see that they are not abused?

41. Can any estimates be given of the number of deeds of covenant in force in 1937–8, 1949–50, 1954–5 and 1959–60 classified by:

(a) income, sex, marital condition and age of the covenantor

(b) number, income, sex, marital condition and age of the covenantees

(c) number and type of covenant and amount of annual payments:

    (i) for payment of sums to an adult other than (iii) (vi) (viii) and (x)

    (ii) for payment of sums to trustees for benefit of infant grandchildren

    (iii) for payment in favour of a discretionary class

    (iv) for payment in favour of a charity

    (v) for payment of sums to infants other than (ii) (vii) (ix) and (x)

    (vi) for payment to married women under separation deeds

    (vii) for payment to infants and children aged 21 and over under separation deeds

    (viii) for payment in favour of parents

    (ix) for payment in favour of infant unmarried children

    (x) for payment in favour of other children

    (xi) for payment of sums by uncles and aunts in favour of nephews and nieces.

42. Do the classification of incomes by ranges tables take account of both sides of these covenant transactions?

43. Estimate of total amount of (a) income tax (b) surtax (c) estate duty forgone in 1959–60 on account of covenants specified under 41.

44. Number of family settlements (including gifts *inter vivos* not settled) in force in 1937–8, 1949–50, 1954–5 and 1959–60 classified by:

(a) income, sex, marital condition and age of the donor

(b) number, income, sex, marital condition and age of the donees

(c) number and type of settlement:

    (i) on infants existing and future

    (ii) on sons and daughters

    (iii) marriage settlements

    (iv) on grandchildren existing and future

    (v) others

    (vi) oral settlements.

45. Estimate of total amount of (a) income tax (b) surtax (c) estate duty forgone in 1959–60 on account of settlements specified under 44.

46. Number of hitherto infant beneficiaries under family settlements claiming repayment of personal reliefs on attaining age 21 or at marriage in 1937–8, 1949–50, 1954–5 and 1959–60. Classified by:

(a) income, sex and marital condition

(b) income, sex, age and relationship of settlor.

47. Number of discretionary trusts in force in 1937–8, 1949–50, 1954–5 and 1959–60 classified by:

(a) income, age, sex and marital condition of the settlor and capital value of the trust

(b) number, income, age, sex and marital condition of the beneficiaries distinguishing:

   (i) children existing and future of the settlor

  (ii) grandchildren existing and future of the settlor

 (iii) other relatives of the settlor

 (iv) employees, directors and their relatives of settlor's company

  (v) all other classes of beneficiaries

 (vi) subscriptions to selected charities

(c) specified period of trust.

48. Estimate of total amount of (a) income tax (b) surtax (c) estate duty forgone in 1959–60 on account of trusts specified under 47.

49. Number of estate duty cases in 1937–8, 1949–50, 1954–5 and 1959–60 which included any covenant, settlement or trust classified by:

(a) size of estate

(b) age, sex, marital condition and country of domicile of deceased.

50. Number of marriage settlements (a) in force and (b) number approved in 1937–8, 1949–50, 1954–5 and 1959–60 classified by:

(a) income, age, sex, marital condition and relationship of the settlor

(b) value of the settlement

(c) number, income, age, sex and relationship of the beneficiaries.

51. Estimate of the total amount of (a) income tax (b) surtax (c) estate duty forgone in 1959–60 on account of settlements specified under 50.

52. Number of trusts, settlements and covenants varied in 1937–8, 1949–50, 1954–5 and 1959–60 and estimate of total amount of (a) income tax (b) surtax and (c) estate duty forgone in 1959–60 on account of such variations.

53. Number of *ex gratia* pensions first paid to (a) employees (b) directors (allowed as expenses or approved by the Board) in 1937–8, 1949–50, 1954–5 and 1959–60 classified by amount of any lump sum paid and by range of income of recipient immediately before retirement:

Lump sum range	*Income Range*
Under £500	Under £1,000
£500– £999	£1,000– £4,999
£1,000–£2,499	£5,000– £9,999
£2,500–£4,999	£10,000–£19,999
£5,000–£9,999	£20,000–£39,999
£10,000–£19,999	£40,000 and over
£20,000–£39,999	
£40,000–£99,999	
£100,000 and over	

54. Number of (a) *ex gratia* widows' pensions and (b) separate widows' pensions for widows of directors first paid (allowed or approved by the Board) in 1937–8, 1949–50, 1954–5 and 1959–60.

55. Estimate of total amount of (a) income tax (b) surtax and (c) estate duty forgone in 1959–60 on account of provisions specified in 53 and 54.

56. Number of trusts embodying pension schemes for salaried directors

and senior employees established by employers and current in 1937–8, 1949–50, 1954–5 and 1959–60 classified by:

  (*a*) number of beneficiaries covered by age and salary ranges
  (*b*) range of lump sums payable as set out in 53.

57. Estimate of total amount of (*a*) income tax (*b*) surtax forgone in 1959–60 on account of trusts specified in 56.

58. Amount of contributions to the following types of pension scheme by employers and employees respectively allowed as expenses for tax purposes in 1959–60 or the latest year of assessment for which information is available in each case:

  (*a*) Funds approved under S. 379 (including the approved past of partially approved funds) distinguishing contributions during the year or spread forward from previous years in respect of past service.
  (*b*) Schemes approved under S. 388.
  (*c*) The unapproved part of S. 379 partially-approved funds approved under S. 388.
  (*d*) Schemes which were in operation before April 6, 1944, and are exempted from the provisions of S. 386 by S. 387 (2) (*a*).
  (*e*) Schemes which were in operation before April 6, 1947, exempted from the provisions of S. 386 by S. 387 (1) (*c*).
  (*f*) The part of the schemes included in (*d*) and (*e*) above which relates to directors or employees admitted to the schemes after the above-mentioned dates.
  (*g*) The part of the schemes included in (*d*) and (*e*) which relates to benefits which can be paid in the form of tax-free lump sums.
  (*h*) Excepted provident funds and staff assurance schemes, as defined in S. 390, (i) in operation before April 6, 1947, and (ii) set up since April 6, 1947.
  (*i*) Statutory superannuation schemes

and number of employees to whom schemes in each of the above groups related.

59. Amount of contributions to S. 379 funds other than ordinary annual contributions made during 1959–60 or the last year for which information is available, which were not allowed as an expense in that year but fell to be spread over one or more subsequent years.

60. Do the estimates of contributions 'allowed as expenses for tax purposes' include contributions by an employer who has made a loss in the year in question? Can any figures be given of the number and amount of such cases in 1959–60?

61. Number of schemes under each heading (*a*) to (*h*) in Question 58 relating to:

1 person
2 persons
3–5 persons
6–10 persons
More than 10 persons

and amount of contributions to such schemes.

Amounts of such contributions relating to:

  (*a*) directors
  (*b*) lump-sum benefits.

62. Amounts paid by employers in last available year for purchase of 'Hancock' annuities, and number of annuities purchased.

63. Number and amount of lump sum benefits classified by lump sum ranges as in Question 53 (other than death benefits) paid to

(a) directors and employees earning £2,000 a year or more and

(b) other employees, in last available year:

  (i) out of S. 379 partially approved funds;

  (ii) under schemes approved under S. 388;

  (iii) statutory schemes;

  (iv) under schemes exempted from the provisions of S. 386 by S. 387 (1) (c) and S. 387 (2) (a) respectively;

  (v) from excepted provident funds etc., as defined in S. 390;

  (vi) under unapproved schemes not included under the above heading;

  (vii) *ex gratia*.

64. Is any information available or obtainable as to the proportions of contributions, benefits or membership of funds or schemes relating to persons in the following income ranges classified by sex:

> Under £500
>
> £500–   £749
> £750–   £999
> £1,000– £4,999
> £5,000– £9,999
> £10,000–£19,999
> £20,000–£39,999
> £40,000 and over

65. What was the total amount of contributions to S. 379 funds returned to employees leaving the employment before retirement age in the latest available year?

66. What was the total amount of contributions to other schemes (excluding those in respect of which the employee had been charged to tax under S.386) returned to employees leaving before pension age in the same year, distinguishing refunds from statutory schemes (taxable under S. 378 (2)) from others?

67. What was the total amount of benefits paid or contributions refunded under superannuation funds or schemes on the death of employees (including the capital value of annuities payable to dependents)

(a) subject to income tax

and (b) not subject to income tax

and what was the additional estate duty charged by reason of such benefits, for the last available year?

68. What are the corresponding figures for employers' life assurance and death benefit schemes not falling within the definition of superannuation funds or schemes?

69. What was the total amount of employers' contributions, actual or notional, treated as income of directors or employees under S. 386 (1) and S. 386 (2) respectively, in the last available year?

70. How many funds or schemes were approved under the Finance Acts

of 1921 (now S. 379) and 1947 (now S. 388) respectively in each year since 1921 and 1947 respectively?

71. From how many such funds or schemes was approval withdrawn in each year?

72. How many such schemes were wound up in each year?

73. How many employees and directors were covered by such schemes at the end of each year since 1921 and 1947 respectively?

74. How was the Phillips Committee's estimate of £100 million as the 'annual sacrifice of revenue' from tax allowances on retirement benefit schemes arrived at (Cmd. 9333, p. 62), and to what year did it relate?

75. Can similar estimates be given of the cost for 1929–30, 1937–8, 1949–50, 1954–5 and 1959–60, with details of the calculations involved?

76. Can similar estimates be given of the cost of life assurance relief for each of the years listed in Question 75, insofar as it is not included in the answers to that question?

77. How many funds are there which are partially approved under S. 379 and under which more than a quarter of the aggregate value of the retirement benefits afforded is payable in lump sum form; in respect of how many employees are contributions currently paid to such funds, classified by the proportion of the benefits payable in lump sum form (e.g. 25 per cent, 50 per cent, 75 per cent, over 75 per cent)?

78. As the definition of income before tax is after deducting superannuation contributions what adjustments are made for those contributions which are treated as life assurance premiums? Can any figures be given of the number and amount of the superannuation contributions on which life assurance relief was given in 1937–8, 1949–50, 1954–5, 1957–8 and 1959–60 and by income range, sex, age and marital status?

R.M.T.
February 1961.

APPENDIX B

1938 Population United Kingdom

	Potential tax units	Total individuals
1. Number of married couples (all ages)	10,852,000[1]	21,704,000[1]
2. Number of single males in Armed Forces and Ships in 1939 (all ages)	530,000	530,000
3. Number of single males 15–19 less assumed 40 per cent in Armed Forces and Ships (212,000). Of balance (1,831,000) 92 per cent assumed in labour force employed and unemployed[2]	1,685,000	1,685,000
4. Number of single males 15–19 not in labour force or Armed Forces (including university population)	—	146,000
5. Number of single males 20+ less 318,000 in Armed Forces and Ships (the balance in total 530,000 single men)	3,939,000	3,939,000
6. Number of widowed males 20+	887,000	887,000
7. Number of males 0–14	—	5,266,000
8. Number of single females 15–19 (1,957,000) of whom 65 per cent assumed in labour force[2]	1,272,000	1,272,000
9. Number of single females 15–19 not in labour force (including university population)	—	685,000
10. Number of single females 20+	4,504,000	4,504,000
11. Number of widowed females 20+	2,227,000	2,227,000
12. Number of females 0–14	—	5,129,000
	25,896,000	47,974,000
Less population increase between mid-1938 and September 1939 (pro rata for tax units)	260,000	480,000
	25,636,000	47,494,000

Percentage of total individuals as potential tax units 54%

Percentage of individuals aged over 14 less those not in labour force aged 14–19 (36,381,000) as potential tax units 70%

[1] Including 480,000 married men (deficiency in National Registration 1939) in population of Armed Forces and Ships in 1939 totalling 1,010,000 for the United Kingdom.

[2] The proportions of 92 per cent and 65 per cent taken from *The Impact of the War on Civilian Consumption* (HMSO, 1945, table XIII-B) after adjustment of age groups from seven to five and exclusive of unemployed and those attending instruction centres.

O

APPENDIX B

1958 Population United Kingdom

	Potential tax units	Total individuals
1. Number of married couples (all ages)	13,134,050[1]	26,268,100[1]
2. Number of single males 15–19 (1,695,000) Percentage in labour force employed and unemployed assumed at 80 per cent[2]	1,356,000	1,356,000
3. Number of single males 15–19 not in labour force (including university population)[3]	—	339,000
4. Number of single males 20+	3,202,000[1]	3,202,000[1]
5. Number of widowed males 20+	928,000[1]	928,000[1]
6. Number of males 0–14	—	6,157,000[1]
7. Number of single females 15–19 (1,556,000). Percentage in labour force employed and unemployed assumed at 73 per cent[2]	1,136,000	1,136,000
8. Number of single females 15–19 not in labour force (including university population)[3]	—	420,000
9. Number of single females 20+	3,211,200[1]	3,211,200[1]
10. Number of widowed females 20+	2,971,800[1]	2,971,800[1]
11. Number of females 0–14	—	5,870,200[1]
	25,939,050	51,859,300

Percentage of total individuals as potential tax units	50%
Percentage of individuals aged over 15 less those not in labour force aged 15–19 (39,073,100) as potential tax units	66%

[1] *Annual Abstract of Statistics*, 1959, tables 11 and 13. The figures for Northern Ireland 1951–8 assume pro rata increases for age and marital condition.

[2] *Op. cit.*, table 15. Adjusted for males from 84 per cent (1951) to 80 per cent (1958) to take account of changes in education population and unemployed juveniles. Similar adjustment for females from 79 per cent (1951) to 73 per cent (1958).

[3] Sources: UGC Reports, *Crowther Report 15–18* (Vol. 1), and Vaizey J., *The Costs of Education*.

APPENDIX C

A Comparison of Personal Incomes by Ranges in the United Kingdom for 1938 as shown by the Board of Inland Revenue and the National Income Blue Book

Range £	BIR No.	BIR £000	NIBB No.	NIBB £000
20,000 and over	2,124	86,000	2,000	87,000
10,000–20,000	5,764	77,000	6,000	76,000
5,000–10,000	17,876	121,000	18,000	123,000
3,000– 5,000	31,351	119,000	33,000	126,000
2,000– 3,000	46,191	112,000	46,000	112,000
1,500– 2,000	50,919	88,000	53,000	90,000
1,000– 1,500	112,448	136,000	130,000	157,000
800– 1,000	93,834	83,000	109,000	96,000
700– 800	73,810	55,000	90,000	67,000
600– 700	108,275	70,000	130,000	84,000
500– 600	176,815	96,000	210,000	114,000
400– 500	315,444	140,000	360,000	160,000
300– 400	710,358	242,000	780,000	267,000
250– 300	811,502	220,000	750,000	204,000
200– 250	1,556,227	345,000	—	—
50– 250	—	—	?	2,700,000
Total above £250	2,556,711	1,645,000	2,717,000	1,763,000

Additional:

160,000   118,000

=160,000 income units with an average income per annum of £738.

APPENDIX D

Life Assurance through Pension Schemes

Tony Lynes

THE income tax allowance for life assurance premiums is restricted in a number of ways, mostly reflecting abuses which have been dealt with by tightening up the law at various times. As a result, the growth of life assurance in its simplest form of the payment of premiums by the individual concerned to an insurance company has been accompanied by the development of a variety of other methods by which the same object can be achieved and which yield a greater saving of income tax and surtax. In particular, advantage has been taken of the tax reliefs given in respect of contributions to pension schemes to provide death benefits either in lump sum or pension form on extremely advantageous terms. The object of this Appendix is to explain the restrictions which have from time to time been imposed on life assurance relief in its simple form and the devices which have been adopted in order to avoid these restrictions. Some illustrative material from the insurance press and other sources will be found in Appendix E.

Gladstone's Income Tax Act of 1853 allowed the whole of the premiums on a life assurance or deferred annuity policy to be deducted from income taxable under Schedules D and E, subject to a maximum deduction of one-sixth of the total income. It is doubtful whether the allowance was intended to apply to pure endowment policies, but the Inland Revenue gave relief on such policies as a concession. In 1913, however, the Court of Appeal held that an endowment policy fell within the scope of the statutory relief.[1]

On the introduction of super-tax in 1910 serious possibilities of tax avoidance through life assurance were opened up. A short term endowment policy could be used to obtain a tax-free lump sum after a few years, the amount of actual life assurance involved being negligible (or none at all in the case of a deferred assurance policy); or a longer term policy could be surrendered for a lump sum—also tax-free. If the super-tax payer did not wish to pay the premiums on such a policy out of his current income, he could borrow the money, and since the tax saving would be greater than the interest charged, he obtained free life assurance and an unearned profit into the bargain.

The Finance Act, 1915, limited the allowable premiums to 7 per cent of the capital sum assured, or £100 in the case of deferred annuity policies. This, however, did not prevent the continued avoidance of tax through the early surrender of endowment policies. Nor did it eliminate the possibility of easy profits through borrowed premiums. The 1916 Finance Act therefore went further and withdrew the super-tax allowance for life assurance premiums completely, limiting the income tax allowance to the standard rate of tax for 1915–16—3s in the pound. As the standard rate was increased for 1916–17 to 5s in the pound, this meant that, in effect, relief

[1] *Gould v. Curtis* (6 TC 293).

would be given on only three-fifths of the allowable premiums. The Finance Bill of 1916 had proposed still more drastic measures. No relief would have been given on endowment or deferred annuity policies under which the capital sum or annuity was to be payable before age 60 or within twenty years from the date of the contract, and tax was to be charged on the sum obtained on the surrender of a policy. The provisions finally enacted, including the restriction of relief to 3s in the pound, were based on proposals made by the Life Offices. With the exception of the withdrawal of super-tax relief, they applied only to policies executed after June 22, 1916.

In 1920 the Royal Commission on the Income Tax recommended that the allowance should be limited to half the standard rate for policies effected after June 22, 1916, and this recommendation was adopted in the Finance Act, 1920. It had no immediate effect as the standard rate was then 6s in the pound. In 1922, however, it was reduced to 5s and it remained below 6s until 1939. From 1940, relief was limited to 3s 6d in the pound, although the standard rate was more than 7s, and since 1948 full relief has been given on two-fifths of the allowable premiums.

The Finance Act, 1930, put a stop to the practice of borrowing the money to pay premiums on a policy, by prohibiting the deduction of the interest from the taxable income for surtax purposes. Without the surtax saving, it was no longer profitable to make this kind of arrangement.

In theory then, life assurance relief is now in 1961 limited to a deduction of two-fifths of premiums which must not exceed in total one-sixth of the taxable income nor, for each policy, 7 per cent of the capital sum assured; and no relief is given against the taxpayer's liability to surtax. In practice, however, relief is obtainable for life assurance premiums much in excess of these limits. One way in which this can be done is for an employer to pay the premiums on his employees' behalf. Since 1947, certain premiums paid by an employer on behalf of his employees are treated as additional taxable income of the employees and attract only life assurance relief on two-fifths of the premiums up to the appropriate limits. Before 1947, however, and to some extent still, the restrictions imposed on life assurance relief by the 1853 Income Tax Act and subsequent legislation do not apply to premiums paid by an employer. He can claim tax relief for the whole of the premiums, and indirectly this advantage can be passed on to the employee in the form of additional remuneration or higher premiums.

By paying part of his remuneration in the form of insurance premiums, the employer can also reduce the employee's surtax liability. While no surtax relief is given on premiums paid by the employee himself, his taxable income can be reduced by the full amount of the premiums if they are paid by the employer and the remuneration reduced by a similar amount.

Another way in which tax reliefs can be obtained in respect of a lump sum payable on the death of an employee is by providing for the repayment of contributions to a pension scheme in the event of death before, or shortly after, retirement. The sum repaid, like the capital sum assured by a life policy, is free of income tax, even if it includes interest which has been accumulated in an untaxed fund.

Finally, provision for an employee's dependants can be made through a provident fund, the employer's contributions to which rank as business expenses for income tax purposes unless they are caught by the 1947 Finance Act (section 386 of Income Tax Act, 1952).

Each of these methods of obtaining tax reliefs on death benefits is subject to certain limitations, imposed by statute or by Inland Revenue practice. The Revenue will not give their approval under the 1921 Finance Act (Income Tax Act, 1952, S. 379) to the part of a pension fund relating to lump sums payable on death which exceed the greater of

    (a) the contributions of employee and employer, plus interest, less any pension already paid, and

    (b) the employee's actuarial interest in the fund.[1]

Within these limits, however, full tax relief (including surtax) will be given on the contributions and on the investment income of the fund. Moreover, under an amendment made by the Finance Act, 1930, a Section 379 fund can provide pensions for the employees' dependants. Strictly speaking, *compulsory* contributions for widows' and orphans' pensions only qualify for life assurance relief, but in practice full relief is allowed under S. 379 if they do not exceed one-quarter of the total contribution.[2] Such pensions are subject to income tax. Nevertheless they represent a spreading of the employee's earnings not only beyond retirement but even beyond the date of his death, involving in most cases a considerable loss to the Exchequer.

If a fund is permitted to pay lump sums on death in excess of the return of contributions or the employee's actuarial interest, it may still obtain *partial* approval under S. 379, provided that no more than a quarter of the benefits is payable in lump sum form, whether on retirement or death. Tax relief will be given on the whole of the employer's contributions to a partially approved fund. Until 1947, there was no such limitation on the proportion of the benefits of a partially approved fund which could be paid in the form of lump sums. The part of the fund relating to pensions could be approved, however small a proportion of the whole it might be. The limit was introduced to bring partially approved funds under S. 379 into line with schemes approved for the purposes of the 1947 Finance Act.[3]

The 1947 Act severely reduced the scope for tax avoidance by the payment of contributions by a body corporate towards tax-free benefits for its directors and employees. Under certain circumstances the contributions are treated as additional income of the beneficiaries. The Act, however, in addition to excluding non-corporate employers, also excluded from the definition of a 'retirement or other benefit' any benefit 'which is to be afforded solely by reason of the death or disability of a person occurring during his service, and for no other reason'. The Board of Inland Revenue explained to the Tucker Committee that in 1947 'it was felt that in the many ordinary staff schemes in which the death benefit was payable only if the employee died before he reached the normal retirement age, there had been no abuse'. The exemption of such schemes from the provisions

[1] Cmd. 9063, 1954, p. 21.

[2] *Ibid.*, p. 17.

[3] *Ibid.*, p. 21.

of the Act, however, provided a loophole which was soon exploited. Since the benefit could be paid on death or disability occurring during service, private companies could—and did—make agreements with their employees and directors under which service was to terminate only on disability or death, thus providing benefits which, although theoretically payable on disability or death, were practically indistinguishable from retirement benefits. Similarly the directors of a company could be appointed for life by the company's articles. There was—and there still is—no limit to the benefits that can be provided in this way for dependants or, in the probable event of disability (however defined) preceding death, to the director or employee himself. 'These devices,' the Board informed the Tucker Committee in 1952, 'are becoming particularly common among controlling directors of private companies, who can thus get around the 1947 Act to the extent of providing tax-free life cover, even though they cannot provide contractual tax-free lump sums on retirement.'[1]

Non-corporate employers—individuals and partnerships—are not driven to such expedients as these in providing death benefits (or retirement benefits) for their employees, since they were not affected by the 1947 legislation. They are free to pay premiums up to any amount on whole life or endowment policies on their employees' lives. The whole of the premiums is deductible from their profits for income tax (including surtax) purposes, and no extra tax liability falls on the employees.

Even for companies the 1947 Act left considerable scope for tax reliefs on schemes giving death benefits, in addition to the exemption referred to above. It is true that a condition of approval under the Act is that the main benefit afforded by a scheme should be a life pension; but for this purpose the life cover provided by an endowment assurance is ignored, unless the benefits are exceptionally large.[2] It is also ignored in calculating whether three-quarters of the benefits are to be paid in pension form—another condition of approval.[3] Moreover the life pension need not be payable only for the life of the employee. A pension to his widow or other dependents is also admitted. The maximum widow's pension which will normally be approved is half the employee's pension, or half the pension he would have received had he survived to retiring age.[4]

The exemption of certain staff assurance schemes and provident funds from the provisions of the 1947 Act left yet another means of providing death benefits. Contributions to such schemes are exempted only if they relate to persons earning £2,000 a year or less at the time when the contribution is made. The contribution made by the body corporate for any individual must not exceed 10 per cent of his remuneration, with a maximum of £100 a year, but this is sufficient to provide a very useful lump sum death benefit.

In addition to the death benefits obtainable from S. 379 funds, schemes approved under the 1947 legislation and schemes set up since 1947 but

[1] Board of Inland Revenue Evidence to Millard Tucker Committee, Document 159.
[2] Cmd. 9063, 1954, pp. 28 and 32.
[3] *Ibid.*, p. 29.
[4] *Ibid.*, p. 32.

exempted from that legislation, there are also numerous schemes which were set up before 1947 and which were allowed not only to retain the tax advantages they already had, but to offer the same advantages to new entrants. As a result, schemes which give death benefits far in excess of anything that would be permitted under the conditions of approval laid down in 1947 are still admitting new members. As the law stands, there is nothing to stop them continuing to do so indefinitely.

Apart from the various kinds of retirement benefit schemes which can also pay death benefits, employers are free to pay *ex gratia* lump sum death benefits up to any amount and to deduct them from their taxable profits in the year of payment. The 1960 Finance Act, which places certain limits on tax-free lump sums payable *ex gratia* on retirement, does not affect lump sum death benefits, on which there is still no limit.

The 1956 Finance Act gave tax allowances to the self-employed, controlling directors and persons in non-pensionable employment, for contributions to pension schemes. Although the benefits must normally emerge in the form of non-commutable and non-assignable pensions, the return of contributions with interest is permitted in case of death before pension age.

APPENDIX E

The use of Life Assurance, Pension Schemes and Trusts for tax avoidance, illustrated by extracts from the Policy Holder Journal *and other sources*

Tony Lynes

LIFE insurance emerged as a significant method of avoiding liability to income tax and surtax (or super-tax as it was then called) during the First World War. In 1914 tax rates rose steeply and the super-tax exemption limit fell from £5,000 a year to £3,000. Life assurance premiums were allowed in full as a deduction from income for tax purposes. Simply by investing substantial sums in short-term endowment policies, a considerable tax saving could be obtained. Under the heading 'Life Policies for the Wealthy—The New Income Taxes' the *Policy Holder*, on February 3, 1915, gave the following example:

	£	s	d
An income of £3,500 a year will be charged:	487	10	0
Assume £500 is invested each year in life assurance premiums, reducing income to £3,000 or below super-tax limit, the payment would be:	375	0	0
Actual net saving:	£112	10	0

In his Budget speech in May 1915, Lloyd George hit at those insurance companies which were deliberately selling endowment assurance as a method of tax avoidance:

> There are indications that efforts are being made to circumvent the Income Tax, and especially the Super-Tax, by means of the development of the endowment policy. A really *bona-fide* endowment policy is a very valuable contribution to the life of this country, but when you have a scheme for an endowment policy for five years, with enormous premiums, that is not really insurance, especially when it is accompanied by circulars pointing out that when you throw in the Income Tax it is really an investment which gives you your money back with 4½ per cent and that the ingredient which enables you to do that is the very heavy Income Tax and the Super-Tax. I think we must put an end to that, and it is very much better it should be done before the thing develops further. We propose, therefore, that there should be an alteration which will discriminate between the *bona-fide* insurance, including endowment, and the mere obvious attempt to evade the Income Tax and the Super-Tax.[1]

By May 1915 the *Policy Holder*, to its credit, had turned its face against this misuse of life assurance, as its comment on the proposal to limit allowable premiums shows:

[1] *Hansard*, H. of C., May 4, 1915, Vol. 71, Col. 1006.

If a man takes out a ten-year endowment policy at an annual premium approaching 10 per cent, he secures a tax abatement, which is disproportionate to the life assurance element in the transaction, and no one can justify the same. If the proposal of the Chancellor means that abatement is only allowed in respect of premiums of £5 per cent or to that extent, every insurance man will say the suggestion is perfectly equitable. Ten-year endowments will be killed to some extent, but life offices are not likely to mourn over that.[1]

The limits actually laid down by the Finance Act, 1915, were more generous than this. Premiums were to be allowed in full up to £100 or 7 per cent of the sum assured. This limitation completely failed to prevent the continued exploitation of life assurance for purposes of tax avoidance. The following year, therefore, sterner measures were taken, including the withdrawal of super-tax relief. The *Policy Holder* again supported the Chancellor's action and attacked the unsavoury practices of certain insurance companies which had rendered such action necessary:

... unfortunately, and to their lasting discredit, many British life offices during the last two years have been a party to the systematic evasion and wholesale exploitation of the income tax by issuing contracts specially designed for the purpose. Two weeks ago Mr McKenna said, 'We feel that the House ought to protect itself against schemes of this kind, which are universally recognized and advertised as schemes for avoiding the income tax.' The Chancellor then added, 'I hold in my hands an example of the kind of thing I have mentioned. It is headed *Income Tax Saved* and continues
   A method which enables you to save every penny of income tax and without recourse to life assurance and at the same time secure an investment at 7½ per cent free from depreciation, offers the best security known to finance, is well worthy of your attention and investigation.'
We can only say that the office responsible for the above has rendered life assurance as a whole a great dis-service. In our opinion the same may be said of the Scottish Amicable, for we find this company, in a special leaflet, drawing particular attention to the fact that 'The privilege of tax abatement is not applicable to any other form of investment, and this is the only method by which a large personal estate can be created by a relatively small initial outlay.'
   We hear of a still more glaring example of tax exploitation. The assured pays the first premium on say a ten year endowment to the company and then claims the income tax rebate. When the rebate is secured the policy holder takes it to the assurance office, the Company arranging a loan to cover any balance of premium number two in excess of the income-tax rebate. This process is repeated for ten years with the result that the policy-holder, having paid one premium only, owes the Company about 70 per cent of the capital sum then payable, the other 30 per cent being the amount he receives in return for the single payment made. This is 'tax dodging' pure and simple, and we are surprised that any old and respectable life office should have been a party thereto.
   'How to Secure the Largest Abatement of Income Tax' is the title of a leaflet issued by another Scotch office which points out in bold type

[1] May 19, 1915, p. 356.

'that with income tax at 3s 6d in the £ the result is practically *a State subsidy to the policy holder of 17½ per cent*'. Whilst this is not so flagrant a case of tax dodging as the one previously mentioned it was bound to attract the attention of the authorities as well as the public.

A British office has a special brochure 'Death and Taxes', explaining 'the only means of securing relief from income tax and super-tax and providing for death duties'. Another asks the question 'Are you confident that you hold a sufficient amount of insurance on your life, having due regard to the relief from payment of income tax and super-tax granted by the Inland Revenue authorities in respect of income applied as insurance premium?' and so on.

We could multiply examples almost *ad lib.*, and can only say we fear that competition has led most, if not all, the companies astray, and that this is perhaps the least creditable chapter in the history of a business which for many generations has been so conducted as to make it an outstanding example of broad minded and high minded administration.[1]

An indication of the amounts being paid by rich individuals in yearly premiums was given in the Report Stage debate on the Finance Bill in 1916. Sir Thomas Whittaker, Liberal MP for Spen Valley and Chairman of the United Kingdom Temperance and General Provident Institution, who might have been expected to have some sympathy for the insurance interest, stated:

> It is the wealthier people who are going to be touched by this restriction of the allowance. They are not the people for whom we need to go out of our way to encourage thrift. We have gentlemen insuring for £100,000, £200,000 and £300,000, and paying £5,000, £10,000 and £15,000 a year in premiums. They are to have an allowance of 8s 6d in the £! I say it is monstrous that they should be relieved of taxation in war-time to that extent. . . .
>
> When this House in 1907 decided to differentiate between earned and unearned income and to levy a lower rate of tax on earned incomes than on unearned incomes the whole reason for this allowance disappeared. . .
>
> Even if you suggest the idea of thrift, Members of this House since the War began have taken up policies for £100,000, and are paying premiums of £5,000 or £6,000 a year.[2]

Paradoxically the most hostile criticism of the insurance companies came not from the supporters of the Bill but from Sir Edward Carson who opposed the limitations to which the life offices had agreed:

> What is the case of the right hon. Gentleman? He says, 'I have settled all this with the insurance companies.' I do not care about the insurance companies. Insurance companies are the most illiberal companies I know.[3]

Although the life offices themselves had played a prominent part in drafting the provisions of the 1916 Finance Act aimed at stopping the use of endowment policies for tax avoidance, they remained on the lookout for new loopholes. Soon after the war the 'single premium policy' became

[1] July 5, 1916, Vol. 34, p. 419.
[2] *Hansard*, H. of C., July 12, 1916, Vol. 84, Cols. 370–2.
[3] *Ibid.*, Col. 367.

popular. It owed its popularity to the fact that the greater part of the single premium could be borrowed from the insurance company. Interest on the loan, unlike annual premiums, attracted surtax relief. The premiums received by the life offices on single premium policies rose from about £600,000 in 1923 to well over £12 million in 1928. Surtax relief on the loan interest was withdrawn in 1930. The boom came to an abrupt end and in 1931 single premiums were down to £2,800,000. The *Policy Holder* was as usual opposed to the exploitation of life assurance for tax purposes. Its life assurance feature writer, 'Oudeis', described twenty-five years later the campaign against single premiums in the 1920's—a campaign in which Frederick Schooling, a director and former joint general manager of the Prudential, played an active part:

> ... just over twenty-five years ago ... Some clever individual hit on the idea of issuing an endowment assurance for a vast sum assured on a single-premium basis, with almost the whole of the single-premium provided by the life office as a loan. Later, the practice developed into using a pure endowment instead of the endowment assurance.
>
> ... the basic contract (ignoring tax influences) often ran at an actual cash loss; yet stories got around of fabulous policies being written on this basis, and there were plenty of men ready to take up such policies.
>
> ... I proceeded to attack the tax evasion plan critically.
>
> The outcome was a very mixed bag of letters. A few men got extremely hot under the collar, and told me in no measured terms how wicked my line of comment was. The majority gave me cordial support. . . .
>
> Then—quite suddenly—the critics went completely silent, because Sir Frederick Schooling took up the theme in the *Daily Telegraph* . . . [in 1925]
>
> Let me quote from his opening broadside . . .
>
> Some little while ago an ingenious insurance man discovered that super-tax payers could make other people pay some of their taxation for them by effecting policies at single premiums and borrowing the greater part of the single premium. They are within their legal rights in doing this, but to my mind it is unseemly, and, incidentally, foolish for life assurance companies to encourage this evasion of tax paying.
>
> The Companies which are definitely pushing these tax evasion policies are doing the best they can to invite Government action for the protection of tax-payers in general. The income tax concessions in connection with life assurance are exceptionally generous for the well-to-do, and if life offices promote a system of evading the manifest intentions of Finance Acts, it would only be appropriate that the Government should withdraw part of the somewhat excessive concessions . . . it may be that experience will prove to some of the tax evaders that they have not been quite so clever as they thought they were being . . .

Oddly enough, in spite of that onslaught and further detailed comments in the press by Sir Frederick Schooling, the practice continued for quite a time, until eventually a Chancellor of the Exchequer dealt with it in a Finance Act—upon which there was a squeal because the men who had been advocating those 'super-tax saving policies' were shocked to find that such a thing as retrospective legislation could exist . . .[1]

[1] *Policy Holder*, May 10, 1950, Vol. 68, p. 451.

Among those who squealed unsuccessfully against the retrospective effect of the 1930 Finance Bill (retrospective only in the sense that it withdrew the surtax allowance from existing contracts) was the *Policy Holder* itself:

> Within the past few years life offices have granted a considerable number of single-premium assurances framed for the express purpose of enabling the policy-holders to reduce their liability to sur-tax . . . It now seems that the present Government, contrary to all precedent, has decided so to frame the portion of the Finance Bill dealing with this matter as to create a definite precedent of retrospective legislation, and we were pleased to note that during the House of Commons debate on May 6th, the matter was taken up by several members of Parliament.
>
> It is most unfortunate that at the present time, with a Socialist Government in office, life offices are not well represented in the House of Commons and particularly unfortunate that those who should have been spokesmen on this occasion did not seize upon the opportunity. The Division Lists show that directors of life offices very intimately concerned with this matter were present, and their silence is to us unaccountable.[1]

The Government used the occasion not only to deal with the single premium racket but also to exact a general undertaking from the Life Offices Association that they would not in future countenance the issue of policies whose main object was tax avoidance. Some months later, commenting on a suggestion that single premium policies had not been effectively eliminated by the Finance Act, the *Policy Holder* put the responsibility for fair dealing squarely on the life offices:

> In our view, it is a point of honour for life offices henceforth to see that no further attempt is made to find a flaw in the technique of the Finance Acts which will enable life assurance to be used as a means of lowering taxpayers' liability to sur-tax.[2]

About the same time a new type of policy was becoming popular—the 'family income policy'. It consisted essentially of a life assurance providing an annuity from the date of death to a fixed future date. Thus, if the policy were for a period of twenty years the annuity would become payable only if death occurred within that period and would cease at the end of the twenty years, when in many cases a large final lump sum would be paid. For a man who wished to provide, for example, a guaranteed income for the education of his children, such a policy was attractive quite apart from the taxation aspect. In addition to its more obvious advantages, however, it had the further attraction that the annuity was tax-free, since it was considered to be a series of capital payments. Despite this fact, the family income policy never seems to have been considered as primarily a tax avoidance scheme, though the tax saving must have accounted for much of its popularity.

It was not long, however, before some of the life offices became aware

[1] May 14, 1930, Vol. 48, pp. 840–1.
[2] October 8, 1930, Vol. 48, p. 1663.

of the possibility of exploiting another scheme which was tax avoidance in its purest form. This was the 'family trust' under which income could be transferred from a parent to his children with a considerable saving of income tax and surtax. The *Policy Holder* was prompt and severe in its condemnation:

> A few years ago the *Policy Holder* criticized strongly a tax evasion scheme operated on the foundation of single premium assurances, and it was then hoped that a long time would elapse before the untimely ingenuity of some clever person should return to the subject. Unfortunately, this has not been the case, and a plan has been devised to enable income tax and sur-tax payers to evade their liabilities by the creation of fictitious trusts for children . . .
> This scheme is being used at the present time to attract the attention of men who are paying a substantial amount of income tax and sur-tax, and is linked with life assurance by the stipulation 'If we are able to secure a repayment of income tax of twenty or thirty pounds on your behalf, will you undertake to effect a life assurance through our agency . . .' . . . like other forms of tax evasion, it cannot do anything but harm to life assurance if it is in any way linked with that beneficent institution, and for this reason we express the wish that all insurance companies shall counsel their officials very definitely against the scheme.
> No doubt some of our readers are already aware of the firm which is using its professional status to add weight to the recommendation of these trusts, and we hope the Law Society will do its share in looking into the matter at an early date.[1]

This article was followed, some weeks later, by a letter to the editor from Messrs Wilkins, Docwra and Company, Chartered Accountants, defending the use of family trusts for tax avoidance:

> . . . granted there is a definite and legally binding transference of a portion of the parent's income to the trustees of the child, such schemes in no way call for criticism, but rather for the approval which eminent taxation lawyers have accorded to any scheme which takes full advantage of the provisions of fiscal legislation.[2]

The Editor was totally unrepentant and reminded the life offices of their earlier undertaking:

> Where a wealthy man would in normal circumstances arrange trusts for certain purposes . . . any gain in the way of tax-saving is perfectly legitimate . . .
> But when a man . . . is invited to set up an absolutely fictitious arrangement solely to enable him to evade the payment of tax to which he is liable, the position is very different . . .
> . . . the 'circular' advised that the father should be one trustee and, apparently, one of the solicitor's clerks could be the other—with the eventual result that there would be no transference of income in any shape or form. That is made thoroughly clear in the circular, and it is certain that the least suggestion of 'transference of income' would kill

[1] October 25, 1933, Vol. 51, p. 1795.
[2] January 17, 1934, Vol. 52, p. 78.

the idea in the majority of cases.

... in recent insurance history there is a black page which records a flagrant attempt to encourage tax-evasion by means of leasehold redemption of sinking fund policies. This was known as the single-premium-cum-loan stunt, and it was freely advocated right up to the day when it was legally crushed. The subject was debated in Parliament, and no one could be found to put up a word of defence, but on the other hand we seem to remember that the Life Offices Association entered into a sort of 'Gentlemen's Agreement' with the Treasury which should exclude the chance of insurance ever again being used for such a purpose.

Unhappily, the circulars in the present tax-evading effort were in part distributed through insurance channels and we are glad to know that after we had drawn attention to the plan it came to an ignominious end.[1]

The *Policy Holder* was apparently over-optimistic in supposing that its disapproval would suffice to kill this lucrative scheme. Two years later the Government was compelled to take action to stop the drain on the Revenue (estimated at £2,500,000 a year) due to what were by then known somewhat euphemistically as 'educational trusts'. Announcing this decision in his Budget speech, Neville Chamberlain described the ways in which the tax avoidance gospel was being spread, by insurance agents among others:

> The other day I came across a printed document which I am told is being issued wholesale all over the country and which gives the most careful directions as to how Income Tax may be avoided by this method. It says, for instance:
>> In our opinion you would meet technical requirements if you draw a cheque payable to Deed of Covenant or Bearer, and pay it into your own Bank Account, as if it were a dividend Warrant. This constitutes a desirable contra entry in your Bank Book, but does not involve an out-of-pocket transaction.
> It concludes by saying:
>> Your friends will thank you for an introduction to our scheme, and I should be happy to send you a cheque for 10s 6d in respect of any new client introduced and accepted by us.
> It is easy to see what is happening. A householder in one of our suburbs receives this document one morning at breakfast and, seeing that it is about Income Tax, of course he reads it with interest, and with growing delight, as he finds that, with perfect safety to himself and in strict accordance with the law, he can substantially reduce his liability to Income Tax. Of course, as soon as he has assimilated this plan, he remembers the precept that you should do to others as you would be done by, and, not despising the half-guinea, he does an act of neighbourly kindness by passing the document over the fence. In due course his neighbour also makes himself acquainted with the plan and earns another 10s 6d and before long it has gone right through the street, and there is not a householder in the neighbourhood who has not been introduced to this philanthropic agent who has come to the assistance of the British taxpayer. But there is by no means a monopoly in this

[1] January 17, 1934, Vol. 52, p. 77.

business. Agents and canvassers are busy hawking about schemes of this kind on behalf of various companies, and they are making them a means of getting insurance business. I have seen another document left by an insurance agent, which says:

> We are going to save you money, and the only condition which I wish to make is that you regard this as confidential and apply every penny of the savings for the benefit of yourself, wife, and children on a Policy with my Company. This would be the only gain I should receive, and as this is my profession, I depend upon your word not to divulge the scheme to any other insurance representative.[1]

Another loophole had been stopped, but the ingenuity of the insurance companies was by no means exhausted. With the War and the introduction of Excess Profits Tax, the rewards of tax avoidance became more attractive than ever. It was at this time that pension schemes, which had always attracted generous tax concessions, began to be used deliberately for purposes of tax avoidance. As a result of a decision by the Board of Referees, employers were for a time able to deduct contributions to a pension scheme in respect of past service from their current profits. Their liability to EPT could be greatly reduced in this way and a spate of large back-service payments to pension funds occurred. The Finance Act, 1943, made such payments less profitable by providing that they must be spread over a number of years. There is no direct evidence as to the role of the life offices in this particular episode, but it seems certain that they were involved to some extent, since the Chancellor of the Exchequer stated that the Inland Revenue had 'consulted those acquainted with the matter in the insurance world' before deciding how to deal with payments of this kind.[2]

At the end of 1943, the *Policy Holder* reported a revival of interest in life policies drawn up so as to avoid aggregation with the main estate of the deceased for purposes of estate duty. This was possible as a result of the provisions of the Married Women's Property Act. An example was quoted of an estate of £410,000 including life policies for £60,000. By writing these policies in the appropriate manner, the duty payable on the estate could, it was claimed, be reduced from £145,000 to £119,000. The *Policy Holder*, however, not only condemned the scheme but expressed doubts as to its effectiveness:

> A professional firm in a northern town (not, we are glad to say, a firm of insurance brokers) has been advising wealthy men that if they effect a life assurance carrying a particular form of clause under the Married Women's Property Act, the proceeds of that life assurance will not be 'aggregated' in the main estate . . . the whole scheme is a tax evasion dodge, and it is completely unsound. We have consulted a leading tax authority on the subject, and he assured us that no matter how the policies were written, they would be treated as 'property passing on death', and would combine to form the aggregate sum of £410,000 quoted in our statement . . . And he added that 'if one estate got away with it, the law would immediately be redrafted to exclude the possibility of a repetition'.[3]

[1] *Hansard*, H. of C., April 21, 1936, Vol. 311, Cols. 46–7.
[2] *Ibid.*, June 2, 1943, Vol. 390, Col. 287.
[3] December 22, 1943, Vol. 61, p. 841.

The 'leading tax authority' turned out to be wrong. The device worked only too well and no attempt was made to amend the law so as to prevent the continued avoidance of estate duty by these means until 1954.[1] Early in 1944, the *Policy Holder* published two letters from readers with experience of the successful use of Married Women's Property Act policies, and a leading article reiterating its opposition to the practice:

> . . . at the present time the whole trouble arises because certain people are actively pushing life assurances containing a special clause, and are arguing openly that in this way the owners of the policies may evade payment of estate duties. When we wrote the first article on this subject, a policy for half-a-million was being hawked round in parcels, subject to the appearance in the document of a particular clause selected by the man who was handling the business. In a second case a wealthy man was advised by letter to sell securities (on which he was receiving a very slight net income) and to use the money by way of life premiums—the final result being demonstrated as a very considerable escape from estate duties by means of the special clause in the policy. The Treasury is bound to come up against these cases, and as soon as it is appreciated that life offices have in any way lent themselves to the practice, much damage will be done to their reputation.[2]

Nor was this abuse confined to a few isolated cases. In an editorial comment following further correspondence on the subject, the *Policy Holder* stated:

> . . . letters from all over the country, from correspondents who claim to have handled many policies of the type in dispute, provide convincing evidence of the scale on which the idea has been practised. Now that evasion of death duties by means of the Married Women's Property Act has been taken up as a 'selling point' by a considerable number of insurance men, why should not the idea become universal . . . ?[3]

Three years later the idea was still being actively promulgated.

> At the present time, efforts are being made in certain quarters to interest wealthy men in life assurance schemes based on clever 'trust' devices, but in our view these have only to attain any considerable proportions to be rendered illegal by the Chancellor of the Exchequer. Beyond the range of such tax-saving dodges, wealthy men do not take up life assurance in this country, and for this position the heavy death duties are largely responsible.[4]

On June 4, 1947, a front-page article by 'Oudeis' attacked the insurance brokers who were pushing these policies:

[1] The Finance Act, 1954, only provided for the aggregation of all policies to the proceeds of which the same person was absolutely entitled. It is therefore no longer possible to avoid estate duty completely by taking out a series of £3,000 policies for the benefit of one's wife. Despite this limitation, the scope for avoidance remains considerable.

[2] January 5, 1944, Vol. 62, p. 11.

[3] January 26, 1944, Vol. 62, p. 65.

[4] *Policy Holder*, January 1, 1947, Vol. 65, p. 5.

P

. . . there are a few insurance brokers whose approach to life assurance is fundamentally wrong. I receive occasionally circulars and other publicity from some of these sources, some of them based on the idea of persuading wealthy men to give up their title to a large sum of money, in return for the possession of an annuity and a string of life policies which purport to replace the whole of the money devoted to this purpose. The scheme turns on what I believe to be an entirely incorrect interpretation of the scope of the death duties, and I wish that the activities of this very minute section of insurance brokers could be promptly checked.[1]

In April 1950, 'Oudeis' reported a move on the part of the life offices to limit the use of life policies for tax avoidance under the Married Women's Property Act.

I have received several advices to the effect that life offices are instructing their representatives not to attempt to arrange more than one policy within any period of twelve months in terms of the Married Women's Property Acts for the benefit of the same beneficiary absolutely. I understand that this applies to all cases where the total sums assured exceed £2,000.[2]

I applaud this step, because it will bring to a close the efforts of certain advocates of life assurance to induce wealthy men to siphon off part of their income (or even capital) to provide premiums for life assurances planned to evade the death duties. It was time that this objectionable cult came to an end.[3]

This provoked an indignant protest from a Glasgow firm of insurance brokers, Messrs Wm. Keir, Bloomer & Co Ltd, who, after defending the practice of paying premiums out of capital on the grounds of 'the small earning capacity of capital in relation to the high demands of surtax', continued:

Your magazine is in its first purpose, a journal for the insurance industry but is available on book stalls to the general public. Such reference as is made by 'Oudeis' to a class of assurance which is willingly accepted by the life offices under Acts on the Statute Book, and clearly within the law of the country, can contribute nothing but bad service to the insurance industry, and as one company of insurance brokers enjoying in this city a good reputation as responsible consultants, we take this occasion to register our protest against such matter being printed in a trade journal.[4]

'Oudeis' remained unrepentant:

. . . there are many life offices determined to bring such arrangements to an end, and I am prepared to argue that before long all life offices will do so. I think that Mr Bloomer will find that a committee will soon be

[1] *Policy Holder*, June 4, 1947, Vol. 65, p. 427.
[2] Estates not exceeding £2,000 were exempt from estate duty. This limit was raised to £3,000 in 1954.
[3] April 19, 1950, Vol. 68, p. 368.
[4] May 3, 1950, Vol. 68, p. 438.

investigating the whole subject of evasion and estate duty . . . Failing some strong arguments from the angle which Mr Bloomer takes up, I very much fear that there will be retrospective legislation covering the contracts which he has arranged. Life offices are anxious to bring to an end all schemes associated with tax evasion of any type—whether they are doing so from their own volition or through Government pressure I am not in a position to say, but the facts are entirely on my side, and tax evasion plans have had their day.[1]

Once again, the *Policy Holder* seems to have been over-optimistic in regard to the future behaviour of the life offices. Even if the 1950 agreement was more effective than the undertaking given by the Life Offices Association twenty years earlier, it was intended (like the provisions of the Finance Act, 1954, which replaced it) to stop only the most extreme abuses of the Married Women's Property Act. The writing of a string of non-aggregable policies for the benefit of one person (normally the spouse of the deceased) was no longer permitted, but even since 1954 it is still possible to take out a number of policies for the benefit of different persons, and no estate duty will be payable on them provided that the sum assured by each policy does not exceed £3,000. A policy for a larger sum will attract duty but if properly drafted will not be aggregated with the rest of the deceased's property, so that a considerable saving of duty may still result.

While the *Policy Holder* was carrying on its intermittent campaign against estate duty avoidance, schemes for avoiding income tax and surtax by means of pensions and lump sum retirement benefits were growing apace. In particular this period saw the phenomenal growth of 'top hat' schemes and their partial defeat by the Finance Act, 1947. On this subject, however (just as, in 1943, on the subject of EPT avoidance through past service contributions), the insurance press was curiously reticent. Apart from a comment on the 1947 Finance Bill, even the ever-vigilant 'Oudeis' remained silent until 1949, when he wrote, referring to an address by Mr John Senter to the Corporation of Insurance Brokers in which he discussed the provisions of the 1947 Act:

As Mr Senter so aptly said 'the sins of the greedy few have led to counter measures that hit the many' and I wish it were possible to write that 'the greedy few' by straining their wits had devised methods of employing retirement benefit schemes which were accepted by insurance companies in complete ignorance of their unwise purposes. But unfortunately the initiative came from certain companies, whose endeavour it was to convince businessmen that 'as the taxation law then stood, they could reap a golden advantage by escaping tax payments on a large scale'.

I write 'certain companies' because there were many whose faces were sternly set against such practices, just as they have opposed every malpractice from the days of the horrible 'single premium plus loan' supertax saving device onwards.[2]

The 1947 Finance Act, however, had not by any means put an end to the

---

[1] May 3, 1950, Vol. 68, p. 438.

[2] *Policy Holder*, February 16, 1949, Vol. 67, p. 119.

use of pension schemes for tax avoidance. In January 1950, in reply to a correspondent who had suggested that 'the present mass of conflicting legislation should be simplified', 'Oudeis' wrote:

> ... why is legislation so complicated? And what are the chances of simplification? It is complex because time and again the Chancellor of the Exchequer has had to step in to check malpractices, and ... there is a practice now in process of extremely active development which may well be the subject of later legislation.
>
> I have before me an example, with details sent to a business friend. He is 60 next birthday; under an existing scheme he is entitled to £500 a year at 65. His salary stands at £3,000.
>
> The offer—pressed by a firm of insurance brokers—is a 'supplementary pension scheme' for this one individual to provide another £1,500 a year at 65. Annual premium for five years £3,500. Offer bearing a late date in December 1949.
>
> If the firm raises this managing director's salary, an outcry will be forthcoming from 'the other ranks'—so, with the aid of a fictitious 'supplementary pension scheme for one', he is to gain by an indirect bonus.[1]

When another correspondent expressed approval of this 'supplementary pension scheme', 'Oudeis' recalled his early campaign against single premium policies:

> Protests ... might have moved me had not famous insurance men (Schooling, for instance) taken steps to ventilate the unpleasant subject in the daily press ...
>
> Unhappily, such efforts have been recurrent. There was one which was killed by the 1949 Budget,[2] and (as usual) the advocates thereof were full of good intentions, and sorely wounded because others (including the Life Offices' Association) took a strongly critical view.
>
> ... a great firm [of brokers] has been circularizing its clients to the effect that this is the sort of scheme which *in the past has invited retrospective legislation.*[3]

Later in the same year the subject cropped up again. 'Oudeis' had again written in critical terms of 'these innocent little "Top Hats" '. In a subsequent issue he quoted a letter from a reader 'possessed of the highest technical qualifications'.

> Your two paragraphs ... seem to leave all the dirt on the Top Hats and their wearers, mainly on their wearers. Is it not the case that if these wearers had never been advised by their various advisers, maybe insurance brokers, maybe chartered accountants, maybe solicitors, maybe insurance salesmen, or maybe even mere commission hunters, they would never have thought up these Top Hat Schemes?

---

[1] *Policy Holder*, January 11, 1950, Vol. 68, p. 27.

[2] Section 25 of the Finance Act, 1949, dealt with an ingenious scheme devised by the Wesleyan and General Assurance Society for the payment of a tax-free annuity disguised as loans secured on a life policy.

[3] *Policy Holder*, January 25, 1950, Vol. 68, p. 94.

Cannot all this restrictive legislation of the past several years be blamed on those wretched advisers . . . ?

I consider the Top Hatted as innocent as newborn babes compared with some of their particularly dynamic advisers.[1]

As always, ingenious arguments were adduced to justify what was plainly a plan for tax avoidance. In 1952 'Oudeis' wrote:

> I have listened several times to advocates of 'Top-hat' schemes whose arguments seemed to follow two main courses—*first*, they maintained that highly paid executives could not build up sufficient capital to buy an adequate pension at retiring age, owing to the inroads of income tax and surtax; and *second*, they argued that if the said executives died, the capital payment would be invaluable in providing for death duties. I have found it somewhat difficult to appreciate how these two conditions could synchronize, and how a man whose estate would need (say) £20,000 for death duties could be in the sad position of lacking the money to buy a pension![2]

At this point our anthology comes to an abrupt and ominous close. There is no evidence that the 1950's were a period during which insurance agents and brokers saw the error of their ways and ceased to exploit the taxation system as a means of selling their wares. And yet the insurance press, so far as appears from an admittedly brief and uncomprehensive survey, abandoned its watchdog role. Even the *Policy Holder*, for nearly four decades a consistent critic of the unethical practices of the industry, was now content to sit on the fence.

[1] *Policy Holder*, November 1, 1950, Vol. 68, p. 1043.
[2] *Ibid.*, April 3, 1952, Vol. 70, p. 307.

APPENDIX F

Number of Incomes at Different Income Ranges 1937–8 and 1957–8[1]

Range of income before tax 1937–8 (1) £	Number of incomes 1937–8 (2)	Range of income before tax 1957–8 on assumption of a two-and-a-half fold rise in personal income per head (3) £	Range of income before tax 1957–8 on assumption of a three-fold rise in personal income per head (4) £	Actual number of incomes 1957–8 on assumption Col. (3) (5)	Actual number of incomes 1957–8 on assumption Col. (4) (6)
200– 499	3,393,531	500– 1,249	600– 1,499	11,513,000[2]	9,060,000
500– 999	452,734	1,250– 2,499	1,500– 2,999	918,000[3]	539,000
1,000–1,999	163,367	2,500– 4,999	3,000– 5,999	194,000[4]	144,000[7]
2,000–3,999	66,493	5,000– 9,999	6,000–11,999	54,000	42,100[8]
4,000–5,999	17,789	10,000–14,999	12,000–17,999	9,150[5]	4,575[9]
6,000–7,999	7,383	15,000–19,999	18,000 +	3,050[6] }	4,125[10]
8,000 +	11,641	20,000 +	—	2,600	
	4,112,938			12,693,800	9,793,800
			Incomes £180–£499	8,105,000	11,005,000
			Incomes £180–£599	20,798,800	20,798,800

[1] Classification of incomes by ranges of income before tax tables (*BIR* 83 and *102*).

[2] 10,620,000 (£500–£999); plus ⅔ of 1,340,000 (£1,000–£1,499) = 893,000.

[3] ⅓ of 1,340,000=447,000; plus 334,000 (£1,500–£1,999); plus ⅔ of 205,000 (£2,000–£2,999)=137,000.

[4] ⅓ of 205,000=68,000; plus 126,000 (£3,000–£4,999).

[5] ¾ of 12,200 (£10,000–£19,999)=9,150.

[6] ¼ of 12,200=3,050.

[7] 126,000 (£3,000–£4,999); plus ⅓ of 54,000 (£5,000–£9,999) = 18,000.

[8] ⅔ of 54,000=36,000; plus ½ of 12,200 (£10,000–£19,999)=6,100.

[9] ¾ of 6,100 (£12,000–£19,999)=4,575.

[10] ¼ of 6,100=1,525; plus 2,600 (£20,000+).

This table is a complicated one because the values for the 1937–8 ranges do not in most cases fit the ranges used by the Board in its 1957–8 table. But whatever proportions are taken for the 'broken' 1957–8 ranges little difference would result. The broad conclusions would stand. Nor would the picture be essentially altered if an increase of three instead of two-and-a-half is accepted as representing the change in personal incomes per head over the twenty years.[1]

Regarded simply as a crude statistical exercise, and accepting the official data without qualification, the table shows a remarkable fall in the number of pre-tax incomes in 1957–8 corresponding in value to the income range £2,000 and over in 1937–8. At the level of £6,000 and over before the war, the number of incomes fell from 19,024 to 5,650 or 4,125 in 1957–8.

[1] *The Economist* estimated that total after-tax personal incomes rose 3·8 times between 1938 and 1959 against a 2·8 times increase in prices and a 10 per cent increase in population (January 14, 1961, p. 111).

APPENDIX G

The Five Income Tax Schedules

Schedule	Source of income	Basis of assessment
A	Ownership of land and buildings (excluding mines, quarries, etc.).	Annual value less deduction for repairs, insurance, etc.
B	Occupation of woodlands, parks, etc.	One-third of annual value.
C	Interest and dividends payable out of public revenue.	Interest etc. paid in year of assessment.
D	(*a*) In the case of UK residents, any property, trade, profession, etc. whether in the UK or elsewhere. (*b*) In the case of non-residents, any property, trade, profession, etc. in the UK. (*c*) Annual income not charged under other schedules and not specially exempted.	Income of the year preceding the year of assessment (with some exceptions); in the case of income from a trade, profession, etc. the normal basis of assessment is the income of the accounting year ended in the preceding year of assessment.
E	Any office, employment or pension.	The emoluments or pension received in the year of assessment.

Note: The above summary is intended only as a rough guide to readers who are unfamiliar with income tax law.

INDEX